original skin

For B, Gabe, and Rory:
Not under my skin, but deep in the heart of me.

original skin

exploring the marvels of the human hide

MARYROSE CUSKELLY

COUNTERPOINT
BERKELEY

This edition has been published by arrangement with Scribe Publications, Australia

Library of Congress Cataloging-in-Publication Data is available.

978-1-58243-739-2

Cover design by Natalya Balnova

Printed in the United States of America

COUNTERPOINT
1919 Fifth Street
Berkeley, CA 94710

www.counterpointpress.com

Distributed by Publishers Group West

10 9 8 7 6 5 4 3 2 1

Contents

original skin

Empresses of the Japanese Bath

I FEEL LIKE I am all skin. It's not just that I am naked, but that the warm air presses up against me like a damp blanket. I scrub my skin with the rough flannel, scraping off dead skin cells and other accumulated detritus until it is pink and smooth. It's as though I am reminding myself of the limits of my existence: where the edges of my body meet the air.

Only minutes ago, I was clothed and on the street, standing outside the humble exterior of the Japanese bathhouse, surrounded by warehouses and sandwich bars in a nondescript street in inner-city Melbourne.

From the moment I step through the door with the prim, handwritten sign 'Strictly non-sexual!' taped to the glass, there is no room for ambiguity: all activity here will be undertaken with decorum. The removal of clothes, and the exhibition of the skin and the body it encases, could be read as an invitation to breach the very boundary displayed. Yet the atmosphere at the bathhouse makes me aware of containment, rather than any messy spilling over or melding of forms.

In Japan, where the communal bath has a long tradition, I imagine group nakedness is no barrier to laughter and the exchange of gossip. Here, in predominantly Anglo Melbourne, where nudity is associated with private intimacy, the tone is discreet, muted, almost solemn.

Over the threshold, I remove my shoes and receive a small stack of towels, a flannel, and cotton pyjamas. Disrobing in the change room, I take the flannel and a towel and enter the tub room. Taps and showerheads on hoses are set below a long mirror; shampoo and body wash are placed on a low, narrow ledge. My body must be scrupulously cleaned before disturbing the pristine waters of the tub. The hard plastic of the small stool that I perch on to perform my ablutions presses into my skin.

There are four of us in the women's area, and it seems to me we move carefully and deliberately, as if the ritual of the public bath has bestowed a gravity to our movements and bearing that evaded us while clothed. It is a little like being in church: the muted tones of our voices, low and respectful; the dimly lit space and the air of contemplation.

I lower myself carefully into the water so that I only create the most well-mannered of ripples across its surface—splashing is definitely not an appropriate response to the expanse of dark fluidity. My skin registers the rising tide of hot liquid, and the silken touch of the water soothes not just my body but my endlessly flapping thoughts. Languorous and heavy, my mind, too, allows itself to be cradled in the buoyant embrace of the bath.

Nudity accentuates the body's physical borders. The skin unclothed registers the limits of the body's perimeter more so than when it is swathed in garments. Clothes tend to blur the boundaries between the physical world and the body so that it is more integrated with the medium in which it is enveloped. Here in the bath, I feel the shape that the contours of my flesh leave in the water like a mould cast in silicone.

It's not quite polite to stare at others when they're nude, and yet here we all are in our skin. Being naked in the presence of other nude women is somehow reassuring. One is encouraged to view the body in its entirety—not parcelled up into unsatisfactory bits presented for consumption. The body dressed only in the skin has a wholeness, an integrity that is sometimes kinder than clothes that shield and reveal our shape.

Perhaps I would be less well-disposed towards nudity if some young Amazon—all long legs, high breasts and flat, un-creased belly—was sitting beside me in the weltering heat. But the young Amazons are elsewhere today, and it is just us ordinary women with our dimpled thighs, our puckered stomachs, our unremarkable beauty, moving from the plastic stools to the bath, the sauna, and the change room with deliberate grace and gravity, clothed only in our skins; empresses in our new clothes.

Skin: the body's envelope

'You gotta have skin,
All you really need is skin
Skin's the thing that if you got it outside
It helps keep your insides in'
— 'Skin', Allan Sherman

ONCE, on a long-distance flight, I sat next to a young man from southern Italy. After a brief conversation about how little Sorrento on Victoria's Mornington Peninsula resembled Sorrento, Campania, he closed his eyes and slept throughout most of the 11-hour flight.

I didn't begrudge his silent slumber, however, as it gave me the opportunity to contemplate the delicious milky-brown skin at the nape of his neck. His soft black hair was short, and the delicate groove that ran from the base of his skull to his second or third vertebra was tantalisingly visible. I didn't run my finger along that silky hollow, but I indulged myself in imagining doing it. I was content simply to gaze as he, completely oblivious to the effortless beauty of his own skin, slept on.

OUR SKIN fixes the boundaries of our physical selves and separates us from the rest of the world. It is the wrapping on the package of flesh, blood, and bone that is our body. Pliant, elastic, able to heal itself, it is where we end and the rest of the cosmos begins.

Exquisitely sensitive and able to register the faintest nuances of atmospheric change, our skin lays us bare to a constant bombardment of sensation.

Through our skin we contact the world; with it we touch and are touched. The skin alerts us to texture, temperature, pressure, pain and pleasure. It is scratched, kneaded, rubbed, and pinched and in response is soothed, stung, and irritated, along with our emotions. Demanding to be stroked and massaged, it flushes and blushes, tickles and tingles, itches and burns. Exterior stimuli prompt it to exude sweat and other fluids from its pores and glands. Freckles, dimples, wrinkles, scars, stretch marks, and moles occur like features on a landscape and the skin itself can range in colour from the milkiest white to an intense blue-black. Eruptions of boils, shingles, or pimples may mar its surface, causing pain and embarrassment. Alarmingly, it bruises and bleeds. Blistering and flaking, puckering and stretching—and feeling, always feeling—the skin is in a constant state of response, alerting the body to the conditions that surround it.

Our skin is unique to each of us; not even identical twins will share the same fingerprint. The pattern of whorls, ridges, and lines found on the tips of the fingers are peculiar to every individual, and can be used to identify us or determine where we have been and what we have touched.

If the skin can leave traces alluding to past events, it is tempting to believe that it can also reveal intimations of the future. If you gaze at your palm for long enough it can come to look like a map, or even a landscape; perhaps the mouth of a mighty river delta, or an aerial view of channel country in flood.

The romance of having your fortune mapped out in the contours and swirls of your palm, the etched lines crisscrossing the fleshy pad and revealing the key to your fortune, is seductive. It's an intoxicating idea that you carry the secrets of your life cradled in your hand and yet concealed, only those versed in the lore of chiromancy able to unlock the secrets there.

The scene is easy to conjure up: the wrinkled crone, swathed in appropriately bohemian garb, firmly grasping your hand in hers. She turns your palm upward and traces her finger (be-ringed and grubby) along the lines and creases there. Lovers, children, spouses, health, wealth and contentment: all the veiled secrets of your future are revealed by this inheritor of Romany wisdom and relayed to you when you cross the gypsy's own dusky palm with silver.

The palm—owing to its sensitivity, and because it can be hidden by a curl of the fingers and therefore also be exposed—is a locus for both vulnerability and a sort of sacredness. It is here that the stigmata, the marks resembling bloody wounds which mimic those inflicted by the nails that fastened Christ to the cross, generally occur. Whether you believe stigmata expose a charlatan or someone experiencing intense identification with the passion of Christ, in reality, those iron spikes were probably driven through the base of the hand in order to support the dragging weight of an adult male, rather than through the palm.

Nevertheless, the palm remains a site of nuance. We hold up our hands with the palms facing outwards as a sign of submission or surrender, but also paradoxically as a sign of dominance, defiance, or triumph.

Even among those impervious to romance and dismissive of anything with even a whiff of the paranormal, the skin has a reputation for sensitivity that goes beyond an ability to perceive the physical world. We feel a niggling heat on the back of our neck, and turn to find the eyes of another are trained upon us. When a sudden, unexplained shiver snakes up our back, we explain it by saying someone just walked over our grave.

A similar force was at play when one of the three witches, sensing the approach on a Scottish heath of that soon-to-be slayer of kings and slaughterer of babies, Macbeth, told her sisters, 'By the pricking of my thumbs, Something wicked this way comes.'

Itching skin is an omen—of receiving money or of giving it away, depending on which palm it is that you have to scratch. And whether we count ourselves as superstitious or not, when the skin of our ears burns red we can't but wonder who is talking about us.

Given the centrality of skin to our existence, it is no wonder that our language is rich with allusions to it. In these aphorisms and adages, the qualities of skin deemed inherent to it include its sensitivity, its alignment with the self, and the snugness with which it enfolds us.

'Thin-skinned' is how we describe someone who is overly reactive; the charge is levelled at those who take offence at any perceived slight, acting wounded when no attack or injury was intended. In the same vein, being told you have the 'hide of an elephant' may be taken as a sign of grudging admiration that you are impervious to insults, or as a slight on your insensitive nature.

Something distasteful is said to make our 'skin crawl'. We're all familiar with this sensation: the skin recoiling as if independent of the rest of the body. Shoulders hunch and quiver, prickles of sensation radiate up the back of our neck, and our scalp tightens round the bones of the skull. Often, it's the very things that crawl—spiders, rats, centipedes (or people whom we feel share similar attributes)—that have this effect on our irritable hides.

Someone who has 'saved their own skin' is burdened with the implication that they have abandoned others in order to ensure their own survival. It may be that they leapt from a burning building while others were still struggling to reach the window; or perhaps they disassociated themselves from a failed venture in the workplace. In either scenario, the insinuation is that they have been less than valiant.

People are admired for being in touch with their feelings—able to not only identify their emotions but to express them honestly. My mother would threaten to take payment for a favour 'out of my hide', a nod to the idea that the skin has currency.

Similarly, to 'have skin in the game' is to risk your own money in a business venture. Skin is a tenuously thin membrane, and so a goal narrowly achieved is done 'by the skin of your teeth', and a 'skinflint' is so mean he would try to fleece the non-existent integument from a piece of stone. We might pronounce that 'beauty is only skin deep' in an attempt to counter praise of a gorgeous individual, casting doubt on the calibre of their character.

'Shedding one's skin' is both an image of growth and of re-invention, and yet skin, quixotically, is also a symbol of intransigence. The rhetorical question posed in Jeremiah 13:23, 'Can the Ethiopian change his skin, or the leopard change his spots?' makes the point that one accustomed to doing evil is unlikely to alter their behaviour and do good.

To be comfortable in your own unchangeable skin is to be at ease with the person you have become: your metaphorical skin 'fits like a glove', and you wear your imperfections and idiosyncrasies lightly.

Someone filled with vigour may be described as 'fit to jump out of her skin', her energy almost unable to be contained by the constrictions of her close-fitting epidermis.

Someone who 'gets under our skin' may do so in a pleasurable or abrasive fashion. Regardless of whether we're drawn to or repelled by such a person, their existence provokes a reaction akin to that of a splinter. Impossible to ignore, we must poke and worry at the source of irritation, raking at it with our nails until it's dislodged.

Most crucially, for all its symbolism and associated imagery, the skin, made up of layers and stretched over the entire body, is our body's largest organ. It exists in contrast with the more visceral images of pulsing wet muscles and the red masses of heart, liver, and kidney that usually spring to mind when we contemplate our organs. These are hidden and slightly repulsive, glistening dangerously, revealed only when the body itself is laid open. The skin, as an ideal, is smooth and pliant, inviting connection, promising containment, and defining beauty.

An adult's skin-surface area will measure between one-and-half to two square metres, and be between one and two millimetres deep. Contrast that with the whale shark, whose mighty epidermis complements its massive size with a depth of around 102 millimetres. Visualise your own skin as a pelt with thickets of hair erupting from the head, the armpits, and the groin, stretched out as a rug on a floor or pinned to a wall, much like a trophy hunter might display the hide of a tiger. True, it may not boast the exotic stripes and wild, elemental beauty of the coat of a big cat, but it is impressive nonetheless.

Thickest on the palms of our hands and on the soles of our feet and thinnest on our eyelids, our skin is constantly being rubbed off and replaced by cells that migrate from the deeper layers of the epidermis. Unlike most mammals, we are relatively glabrous, or hairless, which allows for more efficient evaporation of our sweat, and so assists in the regulation of our body temperature. Embedded in the skin, and growing through it, are hair, nails, and sweat glands. Buried within it are blood and lymph vessels, nerve endings, sebaceous glands, and tiny involuntary muscles attached to our hair follicles.

Our skin envelops us, acting as a barrier against invading microbes and chemical irritants. It protects the underlying tissue from injury and infection, helps to regulate the body's temperature, and alerts the body to environmental factors through its nerve endings: too hot, too cold, too toxic, too sharp—the skin alerts us to the dangers, and the comforts that surround us.

Surprisingly tough yet vulnerable, the skin is a frail and all-too-penetrable veil: blades can slice it, fire can burn it, and toxic substances can be absorbed through it. The loss of a substantial amount of our elastic armour will kill us, rendering us unable to regulate our temperature or block bacteria intent on colonising the warm, wet recesses of our susceptible body. Breach it and we bleed.

You only have to see the pulse beneath a baby's fontanelle, where the skull bones are yet to fuse, or the subtle but continuous beat of blood coursing through the carotid arteries at the throat to have the vulnerability of the body's first line of defence impressed upon you. The merest nick of knife—yes, it would have to be well placed—could see our life drain away, leaving nothing but a dry husk.

Because of this vulnerability, is it any wonder then that few of us ever really feel comfortable just in our own skins? We spend most of our lives draped in clothes; poor, naked, hairless apes we are without them.

The skin is a border, and one that is usually heavily protected and shielded from view, and from the elements. Even the triangles of skimpy bathing suits worn on beaches lend a modicum of defence. Not only does one feel less exposed in the obvious way when one is clothed, but also more veiled at a metaphysical level.

Visually, as well as in a tactile sense, our edges are blunted and muffled when we are dressed. Naturalists are a minority, although most of us have dared the pleasure of skinny-dipping at some point in our lives. There is an abandon associated with nudity, a reckless joy exhibited by streakers at the cricket and by the whoops of bathers plunging naked into the ocean.

Of course, it is also the flagrant display of sexual organs that excites and titillates, not simply the unimpeded view of the skin. Still, it is slightly bewildering that public nudity is viewed as anarchic and an effrontery that warrants being arrested.

I didn't brave the cold and dark on the day that the photographer Spencer Tunick came to Melbourne in 2001 to photograph the bare bodies of the city's citizens as they lay on Princess Bridge. Tunick is famous for his photos, taken all over the world, of groups of humanity in the altogether, lying on city streets or standing in orderly, terraced rows along roads, crowding the upward curve of a pedestrian bridge, or lying on their sides before the looming bulk of an enormous ship. In Tunick's images, the variations of skin colour follow the curves and hollows of the bodies in stark opposition to the unyielding urban surfaces that they are often juxtaposed against.

Watching an edited video of the Melbourne shoot on

YouTube, I found it inexplicably moving to see the mass of people—thousands of them—bare-arsed and happily excited in the muted dawn light, stampeding past the ladder on which the artist was perched with his loudhailer. They dropped onto the cold, wet road at his shouted instruction, but before the photo could be taken a man (fully clothed) ran into the shot bearing a large handwritten sign that echoed the words he chanted, 'All men will bow to the name of Jesus Christ.'

'God sent us into the world naked,' one of the participants shouted back as the police dragged the protester away. His remonstrations were akin to objecting to dancing on a Sunday: there was nothing less lewd than this crowd of adults in their birthday suits, grinning like children at a birthday party. I experienced a mild pang of regret that I hadn't dropped my daks, and the rest of my gear, to pose stark-naked with the lot of them: solemn, ridiculous, exposed.

The exposure of nakedness is something we sometimes crave because of the intimacy that it can help us to forge with another person. When we take a lover, our most urgent impulse is to caress our beloved's skin. We delight in the warmth of their body against ours, and explore their skin as if it were a wondrous new terrain. We seek to discover the blemishes as well as the beauty, eager to know their physical shell in intimate detail. We might pause in delight at their mouth, tracing the shape of their lips, marvelling at the difference in colour and texture. Gently, we circle their nipples with our tongues, smiling with delight as these highly sensitive areas of skin tighten and become erect.

According to our predilections, it may be the smoothness of the underside of our lover's upper arm, the hairiness of other parts, or the contrast between the two that intoxicates us. In our desire to get even closer, we attempt to penetrate the barrier of the skin through deep kissing and sexual intercourse so that, literally and figuratively, our bodies are joined.

What would sex be if not for touch? At its most fundamental, sex is, after all, just rubbing your skin against someone else's. The platonic idea of love is all very well, but who would give up the delicious sensation of sinking into another's arms and feeling wholly embraced within their skin?

And it's not simply the feel, but also the smell of another's skin that can transport us. I can recall vividly the sweet sweatiness, completely devoid of staleness, exuded by a young man whom I studied with decades ago. On hot Brisbane mornings, he would arrive at college, having ridden his bicycle up the myriad hills of the western suburbs, and arrive in time for our first class of the day, wet and glistening, and smelling divine.

The touch of skin on skin is not just for lovers, of course. It is a delightful sensation at any age, and is essential to our physical and psychological development. Anxious parents of premature babies huddle beside humidicribs, gently reaching inside to stroke their tiny offspring's bodies. Desperate to hold their babies, but prevented from doing so by the medical paraphernalia, they caress their infant in any way they can.

Babies denied touch at this early stage of life have lower growth rates and spend more time crying than babies who are touched. In most hospitals where pre-term babies are cared for, 'kangaroo care' is encouraged where possible. This involves placing the babies against their parents' chests, inside the parent's clothing, so that parent and baby are skin to skin. Not only does this help to calm the tiny newborns and promote development, it also helps mothers to feel more bonded to their babies, and to express more breast milk. In an article that I read promoting the benefits of such care for all newborns, not just those born prematurely, the baby, its skin pressed up against its mother's, was described as being in its 'natural habitat'.

As a toddler, my youngest son adored the feel of skin on skin. If he caught me having a quick siesta on the couch, he would leap on me, pulling up first my shirt and then his own. Smiling as he lay against me, the warmth of our two skins as they touched was always surprising and vital. I took an almost-guilty pleasure in the sensuousness of it. His skin was wondrous to me. So smooth, so even; I would find myself reaching out to touch it, stroke it, kiss it. It fitted him so perfectly, without a wrinkle; nowhere did it sag or pouch. His young, firm flesh pushed out against his velvety covering, which wrapped itself around him in a taut, tight embrace.

Our skin is the most outwardly reliable indicator of our age. As we grow older, our collagen fibres gradually lose their ability to bind water—a property that gives the skin its elasticity. As a result, wrinkles begin to proliferate. The skin also thins, and often lesions develop as a result of exposure to the sun. We can all look forward to the papery skin of old age, sagging and bagging, more prone to tears and breaches.

In the West, where youth has become fetishised to an alarming degree, an entire industry is based on the desire to keep our skins as pristine as possible. We buy sunscreen to prevent damage, moisturisers to promote elasticity, and lather on anti-ageing creams in an attempt to repair the ravages of time. Paralysing toxins are injected into facial muscles to inhibit frowning and so display an unlined brow. We subject our skin to chemical peels in order to slough off outer layers and reveal newer skin cells beneath. In response to advertising that encourages dissatisfaction with our appearance, we lighten our skin or darken it with cosmetics and UV lamps.

According to fashion or custom, skin is daubed in colour or designs. Tattoos, piercings, and scarification are used as adornment, as a sign of cultural allegiance, or to hint at times of boredom spent in a correctional institution. We tussle with our skin, hide it, drape it, reveal it, or dab it with perfumes and deodorant in an attempt to camouflage our scent. We apply colour to our lips and faces for dramatic effect, striving to project the image of an ideal self—youthful, healthy, and confident—or to show that we are conforming to, or rebelling against, the expectations of society.

Yet our skin is not always a reliable ally—it can betray us, sending signals that belie our words, actions, or attitude. Apprehension, stress, or sexual attraction may cause our sweat glands to seep a watery fluid containing urea, minerals and amino acids. Glands in our armpits, on our faces, the mons pubis, nipples, and the scrotum ooze a milky, viscous liquid in response to emotional stimulus. A rush of blood to the face can signal our shame, arousal, embarrassment, or the fact that we have just told a lie. The cold sweat of fear, and the accompanying odours that spring unbidden from our pores, may alert a foe to our terror and give courage to their assault.

The skin is like a neon light, flashing signals that provide clues to our health, wealth, race, and occupation, and encouraging others to make assumptions about us. A tan line may define us as an outdoor worker, a stretch mark as a mother. In Aboriginal cultures, a boy's transition to manhood might be marked physically on the body through scarification, while in India the shade of your skin might indicate your caste. Tightness around the temples may hint at a facelift, while a certain type of lesion on the skin known as Kaposi's sarcoma could mark you as having AIDS. Strangers assess your skin and use the information found there to make judgements about your morals, your intelligence, and your worth.

In a number of cultural and religious traditions, one piece of the skin on the male body is deemed superfluous or even unclean: circumcision surgically removes the prepuce, or foreskin, the fold of skin that covers the head of the penis. The operation is usually performed soon after birth, although in some cultures it is done much later.

In recent years, circumcision has fallen from favour, deemed to be an unnecessary and painful operation. However, debate continues within the medical profession about the health benefits of removing this relatively small amount of skin, with some research suggesting that circumcised men are less likely to become infected with HIV. Although male circumcision is a much less radical operation than the mutilation of female circumcision, there are still men who feel the loss of their severed foreskin keenly, and who seek to restore it through both surgical and non-surgical methods.

Jesus, a Jew, was circumcised. According to Christian tradition, Jesus ascended into heaven after his death, taking his body with him. This left his foreskin as the only relic of his physical body remaining on earth. At least 13 churches worldwide claim still to have this tiny relic. In one re-telling of the story of St Catherine of Sienna, she has a vision of marrying Jesus in a mystical union where he places a ring made from his foreskin on her finger. St Angela of Blannbekin, an Austrian saint who lived around the turn of the 14th century, claimed to have had a vision where she swallowed Christ's foreskin. Apparently, it tasted intensely sweet.

Like the air we breathe, we take our skin for granted. We seldom remark upon it except in the context of our collective phobia of ageing, fuelled and exploited by the cosmetics industry. Yet it is remarkable; it mitigates and ameliorates the sometimes-harsh world we dwell in, and is at the interface of so much of what we encounter. It is our border, the edge of ourselves, the point where we meet our universe.

I AM NOT AN ACADEMIC, a scientist, a doctor, or a cultural theorist. I am a writer who has become intrigued and captivated by the precariously thin veil of our epidermis, and how it mediates and facilitates our experience of the world.

This is by no means an attempt to write the definitive book on skin. Instead, it is a dalliance with that which wraps us up—a teasing out of some of the gruesome, visceral, personal, and interesting aspects of the human skin.

If I had to pinpoint one moment when the subject of skin became an obsession, I would not be able to do it. It was a gradual awareness that began perhaps when my children were born and I discovered, somewhat to my surprise, that it was their physical selves that I first bonded with. Our skins were our initial meeting place, and it is their enveloping skin that continues to enthral me. This fascination has, in part, been the starting point for my manuscript.

Once I embarked on the task of writing about skin, I came across newspaper and magazine articles almost on a weekly basis that referenced skin in some way: medical breakthroughs, streakers at the cricket, the horrific scarring of those injured in the Bali bombings, a footballer sledging an opponent about his tattoo, instances of racism, alarm about the dangers of tanning salons, and makeup tips of the rich and beautiful. In the face of such a barrage of topics into which to delve and research, I toyed with the idea of simply making lists: the top ten functions of the skin; aphorisms and literary quotes regarding skin; skin conditions you wouldn't wish on your worst enemy; the five ugliest tattoos I've ever seen ... But it became apparent that this would be a far too glib approach, especially when I began talking to people whose skin—or for whom the skin of others—loomed large in their life, their work, their emotions, or their art. For tattooists, burns surgeons, dermatologists, and those whose bodies wear the marks of the sustained assault of fire or disease, the skin is no small thing; it is integral to who they are and how they experience the world. This is true for all of us, of course, whether we are aware of it or not.

In the daily routine of human transactions—small and large, commercial or otherwise—our skins are the interface between us. Not only are our minds, through our skin, 'brought into relation with external objects', as described in that classic text *Gray's Anatomy*, but, on a deeper level, it is through our skin that we connect to our world and the others with whom we share the planet. Apart from intestinal rumblings and the odd heart palpitation, the mind is aware of the body primarily through the skin.

When we touch another person—no matter how fleeting the contact—a bond is established or modulated to some degree: a perfunctory kiss between spouses; a passionate, full-body embrace in the pre-dawn; an overwhelming rush of love as a parent nuzzles their child's neck; a press of the hand from a trusted friend. Other, more casual contacts also occur: a pat on the shoulder, a shake of the hand, a touching of palms as coins are exchanged or a credit card is proffered. Along the way, items are bought, deals finalised, respect given, emotions communicated, attention sought, and pain inflicted.

Skin, as burns surgeon Fiona Wood remarked to me, is 'not just a plastic bag to keep our giblets in'. No, indeed: the skin is a parchment, a canvas, a prison, a barrier, a conduit, a revelation; like silk, like sandpaper, like flaky pastry; the colour of milk, of chocolate, of wheat, of burnt biscuits. As we go about our daily lives, we constantly brush up against our fellow human beings. Each of these individuals exists inside their own tenuous envelope: the only thing that separates them from the rest of the universe—their skin.

Touch Me

> 'She reminded him of the pleasure of being scratched, her fingernails in circles raking his back.'
>
> — *The English Patient*, Michael Ondaatje

THE SKIN has a memory all of its own.

'I remember that touch,' a lover once said to me, recalling the first time I laid my hand on his arm. That small contact, weeks after we had first met, was enough to prompt him to ask me out. Some 18 years later, I still remember it, too, and can see the scene played out in the dingy foyer of the community theatre where I was performing in a play: the liverish red of the foyer walls, steam rising from the tea urn, a cluster of audience members and my cast mates chatting after the show.

'You were good,' he says to me, 'which was a relief, because I didn't know what I'd say to you if you weren't.' I laugh at his honesty and, made brave by the compliment, reach out, the fingertips of my right hand brushing his left elbow. It is enough to confirm the flick of sexual attraction between us. It's what I intended to convey when I extended my hand in that first crucial touch that seared itself into both our skins.

APART FROM ITS MOST obvious purpose—keeping the body whole and integrated—the skin's primary function is that of a sense organ. It is here, in the pliable, vulnerable, elastic skin, that our sense of touch is located. Known as the mother of all our senses, touch is the first to develop and the last to leave us. While our other senses—sight, smell, taste, and hearing—are located in discrete organs, touch is dispersed throughout the surface of our body in the skin.

When we distrust what our other senses are telling us, it is touch that we ultimately rely on to verify our experiences. That lingerie may look beautiful, but only when it is against your skin will you be able to verify whether it is silk. You may hear your beloved's professions of love over the telephone, but until he is in your arms you will be not be completely reassured that it is to you he is still devoted. Despite clearly remembering that I put my passport in my bag, as I approach the immigration counter at the airport I repeatedly search it out with my fingertips to reassure myself that I have not forgotten it.

RIDING INTO THE CITY on my bicycle, a light, feathery sensation alerts me to something brushing against the back of my right hand. I glance down in response to where my grip encircles the handlebar. Before I even fully register what has caused the nerve endings in my skin to alert my brain to something foreign touching me, I flick off a large huntsman spider skittering across the back of my hand. I gasp, and an expletive rushes out as my sharp intake of breath reverses. The spider must have crawled into the hollow inside the handlebar, confident the small, dark place would be an ideal den. The vibrations travelling through the tyres and up through the metal frame of the bike had disturbed it.

My neck stiffens, the skin across my shoulders and down my back quivering as if the spider had run down my spine. Twenty minutes later, a vague itch in the exact spot where the spider had been keeps me glancing nervously at my hand.

TOUCH CAN BRING US back into our bodies in a way that our other senses do not. While breathing in the heady scent of an old-fashioned rose or taking in a panoramic view of the landscape tend to transport us, swelling the boundaries of our bodies to take in that which is beyond us, our sense of touch shrinks us back into ourselves. Stressing our inescapably corporeal existence, touch reminds us of our body's borders more rapidly and completely than any of our other senses. Touch is impossible to escape: we can close our eyes, stuff silicon plugs into our ears, hold our nose, and refuse to eat, but our skin is always receiving signals, impossible to turn off.

So central to our lives is our sense of touch, so intrinsic is it to the way in which we experience the world, that we can barely conceive of a life without it. Loss of the other senses is easier to imagine. Most children play variations of Blind Man's Bluff; earplugs can offer us some insight into a world without sound; and anyone who has endured a heavy cold knows what it's like not to be able to smell—but a life without tactility? How would we know where we ended and where everything else began, if not for touch? Would it be possible to learn to navigate our body's way in the physical environment without the information we glean through our skin?

Monitoring our bodies and our surroundings would be exponentially more difficult without our sense of touch. Thrown back on to our other senses to bridge the gap in our sensory arsenal, sight would become crucial in gauging the position of our bodies in any particular space. Grasping objects, judging the weather, navigating around potentially injurious surfaces—those with the capacity to cut, to burn, to bruise—would become exercises requiring immense concentration. If the lights suddenly went out when we were standing up, we would simply fall over, not being able to feel the ground beneath our feet.

Surely comfort would be much more difficult to give and to receive without touch? Not even sex would make sense. Yes, our lover's beauty would still exist, as would the smell of their hair and the taste of their lips, but without touch I imagine sex reduced to a mechanical task undertaken purely to ensure the continuation of the species. It would be akin to having a shower in order to stay clean, but without the simple, sensual pleasure of warm water on skin.

The gratification we draw from eating would be radically diminished, too, with the texture of food a mystery to our lips, tongues, and fingers. Perhaps our sense of taste would become much more acute to compensate, and we would find that the fresh acidity of limes, the bite of a salt crystal on our tongues, or the sweetness contained in a teaspoon of honey would become even more piquant.

I don't want to underestimate the power and the importance of our other senses, or to wish them away, yet I can only imagine what strange, constricted creatures we humans would have evolved to be without the ability to physically feel our environment. The trade-off for the release from physical pain would be a harsh price to pay for foregoing the kiss of wind on our skin, the subtle pleasure of clean sheets, or the shock of cold water on a hot summer's day, not to mention the more obvious delights of the caress of a lover's touch. It would be impossible to compile a complete catalogue of the sensual delights that would be denied us without our sense of touch, so varied, subtle, and enmeshed it is in our experience of being alive.

We only have to see an infant mouthing every object it can get its hands on to recognise the importance of touch in gaining information about the world. How would babies learn, when still too young to have language or to understand or interpret what they see? Without a sense of touch to immediately signal that they had injured themselves, for example, few would make it to adulthood with all their limbs intact. Surely, too, the moment when an infant becomes aware of its own skin—the boundary that they first become aware of through their sense of touch—sets the child on the road to the realisation that it is a separate entity: what is inside my skin is me, what is outside is not me.

As I wrote earlier, it was through my children's skin that I first came to know them and, I imagine, how they came to know me.

I'm not sure why this was such a revelation to me. I assumed, I think, that I would look into their eyes and immediately know them as my own. In fact, my attachment to them began with the loveliness of their small bodies, rather than with a recognition of their spiritual selves or their emerging personalities. My impulse was to touch, stroke, and explore them to discover who these brand-new creatures were. In many ways, it was a similar impulse to that felt when we take a new lover.

By knowing their bodies, I came to know them. My sons' enveloping skin was the initial interface between us. This running of my hands over their bodies, over their skin, was the beginning of a deep and abiding connection. I can't help but think of the struggle I would have had to bond with my children if I had had no sense of touch.

I was reminded of the intensity of this connection when I asked a friend how things were going with her new baby. 'We have a secret,' she said of herself and the child, 'we're in love.' I knew instantly what she meant, and could almost feel the rush of milk to my breasts that a baby's searching mouth brings. Other physical sensations, the echoes of which will never completely evaporate, rippled through me: the regular tug-tug as a baby's mouth locks onto the nipple and begins to suck; the flutter of little starfish hands either side of the breast; the feel of tiny fingers hooking into my mouth and, with the first urgency of hunger relieved, small hands wandering, patting my face.

This love affair with my children, and my delight in the perfection of their physical selves, their corporeality, has continued. They have reached an age where they are beginning to struggle against the close physical connection that I still long to have with them. But I remember well when they were young enough to allow me to hold them, to stroke them, to tell them how beautiful I found them, and their enveloping skin continues to enthral me.

Despite this intense bond, I remember being overwhelmed by the constant physical demands of having a new baby. For the first six weeks of his life, my first child had barely been out of my arms: I had bathed him, breastfed him, stroked him, slept with him; I had spent hours and hours touching him.

Once, after a marathon breastfeeding session when my baby would cry if he wasn't attached to the nipple, I recall feeling as if I had this enormous parasite sucking the life out of me. My husband, sensing this — or rather, having it shrilly and tearfully conveyed to him — suggested I go out to a movie with a friend, leaving the baby with him. Not long after I left the house, I was aware of a vague, physical sense of loss. I wasn't missing my child in my head or even in my heart — it was a relief to be away from him — but my skin had come to know his intimately, and that's where this loss resided: in my very pores. My selfish heart had pushed my son aside momentarily, but my skin remained faithful and would not forget him.

OUR LIPS, OUR TONGUES, and our fingertips are areas of skin densely crowded with nerve endings; as a result, they are especially sensitive. We use these parts of our bodies to explore and identify objects in our world.

Beneath the fingertips of someone who is blind, the raised dots of Braille characters transform themselves into words, instructions, stories. With touch, we can test if a cake is baked or the washing dry, and recognise the difference between leather and vinyl, silk and nylon.

A surgeon may rely partly on touch to tell healthy from diseased tissue, use her fingertips to find a wiry vein, or discern where one organ begins and another ends by the difference in texture.

Lips pressed to a child's forehead will sense if he has a temperature. Even with our eyes closed, those same lips puckered in a kiss can recognise the difference between a nipple, a fingertip, or the end of a nose.

We explore objects with touch in a variety of ways that we may not even be aware of. Pick up an unfamiliar object now with your hand. Your fingers will immediately begin a series of actions to help you gain information about it. They will rub across the surface of the object to determine its texture — tracing its edges, finding where the surface dips and swoops, where it protrudes and recedes, its patterns and features. Pressing down upon the surface will allow you to gauge its hardness: is it malleable to the touch or does it remain unyielding to the pressure of your hand? How will you assess its temperature? Most likely, you will simply let your fingertips rest on its surface for a moment to gain this information. You might extend your hand, holding the object away from your body to gauge its weight, and then wrap your hand around it to discover more about its form and volume. To garner more precise knowledge of its shape, your fingers will trace the outline of the object.

THE SENSORY RECEPTORS in our skin are highly specific. Nerve endings only respond to particular stimuli: heat, cold, air movement, pain, pressure, vibration. When a specific sensory receptor is excited, a particular sensation is felt: a burn feels distinct from a scratch, a change in air pressure unlike a change in air temperature.

Chains of neurons connect nerve endings in our skin to the spinal cord or at the brain stem, and onto the cerebral cortex. Information derived from these sensory receptors in our skin travel along these chains. We can then respond accordingly and appropriately: scratching an itch, pulling away our hand from a sharp or hot object, slapping an insect.

Given the sensitivity of our skin, why aren't we in a constant state of irritation at the persistent brush of fibres from our clothes, driven mad by the pressure of our feet on the ground as we walk around, or bowed under of the weight of the atmosphere bearing down on our shoulders? Mercifully, we are kept from the insanity of such unceasing stimulation by the fact that our touch sensors respond to *changes* in stimuli, and not to those that are ever-present.

Our sense of touch is able to alert us to information that extends beyond the mere physical banalities of temperature, texture, and atmospheric pressure. It offers us clues that hint at relatively intangible things, such as how others are feeling about us, their intentions, desires, attitudes—all can be discerned in a touch. Is the person grasping our hand trying to assert their dominance or communicate their submission? Are they demonstrating affection, support, solace, approval, or disgust? Our skin will tell us.

The pressure and duration of touch, in addition to where we are touched, can alarm, excite, irritate, warn, comfort, disgust, or seduce us. City dwellers regularly tolerate the press of flesh from unknown fellow commuters. Unfamiliar backs rest against hips, buttocks brush against the back of a stranger's hand. But in the crowded space of at train a peak hour, it takes merely a slight increase of pressure, a subtle change in intensity where skin meets gaberdine or linen, to alert a woman that someone is taking advantage of the enforced proximity to experience a moment of unshared titillation. Even such a relatively minor example of unwelcome touch can make us feel violated.

Touch connects us all, literally. When we empathise with another's pain or grief, share the excitement of our team winning the grand final, or the triumph of a successful business deal, we are moved to touch each other—a shake of hands, a pat on the shoulder, a quick hug. A friend's eyes fill with tears as they recount a failed relationship, a moment of humiliation, or the death of a parent and, instinctively, we reach out to them. A child comes crying with a skinned knee, and we encircle them in our arms.

Dejected team mates pat each other on the back after a loss, or wrestle each other on the ground in an orgy of self-congratulatory delight. These sometimes casual, fleeting, or impulsive connections affirm our place in the world, that we belong.

Psychologists and other observers of human interaction have long been aware that our touching behaviour is influenced by culture, gender, age, and the relationships between them. How we come to connect physically with others, skin to skin, even in the most perfunctory manner, is determined by a host of different factors that at a conscious level we may only be dimly aware of. Your boss might touch you briefly on the arm to gain your attention, but would you take the same liberty with her? A young man in Indonesia might walk along the street with his friend, their hands touching, but would young heterosexual men in Sydney do the same? A close female friend may drape her arm around your waist, but if she assumed the same liberty with your husband or partner how would you react?

Generally, we spend little time analysing the appropriate way to touch others. It is something that most of us know deep within our bones, and is based on learning that began the moment that we emerged from our mother's body and—depending on the year and the circumstances of our birth—were either placed in her arms, held upside down and smacked on the bottom, or whisked away in the latexed hands of a professional for medical tests or intervention. Based on experience and observation built up over our formative years, we know without thinking how the layers of culture, age, intimacy, status, and gender dictate the appropriate way to touch another person.

Occasionally, though, the way we use touch may be more calculated: at the beginning of a business negotiation, how long will you clasp your colleague's hand when you greet them? Will you grasp their elbow at the same time? To garner a friend's allegiance in a fall-out with another crony, will you use touch to subtly underline the bond between you? As a lone woman in a roomful of men, will you decide to initiate a handshake in order to signal your determination to be taken seriously in the discussion about to take place?

'I'M TOUCHED.' We hear the phrase often when someone is trying to describe how a gesture or a word has moved them, directly connecting the realms of emotion and physical sensation. As a matter of course, we regularly use words with tactile overtones to describe our sentiments and moods: the prickle of irritation, the warmth of a smile, the sting of rejection, the weight of loneliness, the buzz of excitement. Anger is 'hot', kindness is 'warm', and to be ignored is to be 'given the cold shoulder'. Given this linguistic link, would emotions still have the same power if we could not feel the physical world?

Without the ability to name and describe feelings in this tactile way, would emotion wash over us in a deluge of feeling devoid of the nuances that we can categorise so precisely: would anger feel the same as frustration, would sorrow be indistinguishable from ennui, passionate love similar to confusion? Perhaps we would wade through a porridge of feeling rather than dart among the sharply defined mosaic of emotions that we are accustomed to. Such puddles of ill-defined perception would make for much murkier interactions with our fellow humans, and negotiating even the closest of relationships would be a formidable challenge.

THE SKIN, it has been said, is the surface of the brain. Certainly, as anyone who endures an attack of hives or a sudden flare-up of eczema as a result of stress could tell you, the skin and the processes of the mind are not entirely separate.

In fact, there exists at a very fundamental level a link between our skin and our psyche. As the foetus develops in the womb, the skin and the central nervous system, which includes the brain, develop from the same ectodermal origins. The ectoderm is the outermost layer of the three primary layers of the embryo. From our very beginning, then, our mind and our skin are intertwined.

Given this connection on such a basic physiological level, it follows a certain logic that researchers in the field of touch have noted the benefits that massage can have on a variety of psychological disorders. Studies have found that regular massages improve the behaviour of children diagnosed with attention-deficit/hyperactivity disorder (ADHD) and make them feel happier. Massage can help girls with eating disorders develop less distorted body images and improve their eating habits. Other studies have suggested that the aggressive behaviour associated with boys may be due to them being touched less than girls.

American researcher Tiffany Field, director of the Touch Research Institutes at the University of Miami School of Medicine, has proposed that teenage sexual promiscuity and pregnancy are on the increase partly because young people rarely receive appropriate, positive touch from their teachers, sporting coaches, and other adults who shape and influence their lives.

We live within a climate that regards adults touching children with suspicion. Fear of being charged with sexual assault has resulted in teachers being instructed not to touch students, or only to do so within tightly prescribed guidelines. Maintaining concern about sexual abuse is entirely appropriate given the shameful revelation over recent years of systematic abuse of children within various institutions, including within religious and welfare organisations. But at what cost do we starve our young people of positive experiences of being touched?

We know that our bodies and our minds both suffer when we are denied touch. In 1990, after the fall of the dictator Nicolae Ceausescu, distressing images from Romanian orphanages flashed around the world. Stricken children, wizened and undersized, with empty, sunken eyes lay motionless in their cribs. Under-resourced and understaffed, these institutions struggled to care for the children living there. The children languished in their cribs, rarely touched. Their emaciated forms and lustreless eyes demonstrated, in the most graphic and heartbreaking way, the detrimental effects on growth and development of not being touched.

The images prompted a flood of offers to adopt the children abandoned in these orphanages, and many of them found their way to Western countries and new homes. Years later, most of these children were still physically smaller and developmentally delayed compared with other children their age.

CAROL NEWNHAM is a neuropsychologist who works in the Parent–Infant Research Institute at the Austin Hospital in Melbourne. Her work assists parents of preterm babies learn how to interact with their children to overcome the many barriers to their development that often exist. These barriers may result directly from the fact of their premature birth, or from the demands of the care to keep these tiny babies alive.

'The medical intervention that is required for babies born at or below 30 weeks often means that babies are denied contact with their mothers, something we know is essential for their growth and development,' Carol says, obviously engaged by her topic and moved by the plight of the families she deals with. 'In addition to this deprivation are the many, and often painful, procedures the babies must undergo.'

Preterm babies must be kept in isolettes, the heated, plastic boxes that maintain the babies' temperature. They are also routinely subjected to painful medical interventions that may include having their heels pricked and blood squeezed out sometimes several times a day, suction may have to be applied to clear their lungs, IVs inserted into their veins, lumbar punctures performed, and tubes inserted to aid ventilation. All this, according to Carol, can have the unintended, but hardly surprising, result that the babies become frightened of being touched, even by their mothers.

Skin and skin contact has a lot to do with the first months of a healthy baby's life: 'The baby's been *in* a body, and with a full-term baby they then do a lot *on* that mother's body,' Carol explains. For babies born prematurely, this contact, which includes touching and massaging, smelling and tasting, and vestibular (whole body) movement, is severely disrupted. For the parents, too, the need to caress and hold their child is almost overwhelming, but the result of a baby associating touch with pain and stress can turn even the simple task of changing a nappy, or giving the baby a bath, into a heartbreaking clash of conflicting needs, high anxiety, unread signals, and tears.

Many of us assume when we become parents that we will intuitively know how to touch our babies, to give them comfort, to soothe them when they cry. The mother–infant dance is what Carol and her team call that 'happy state', when everything falls into place and mother and child react and respond to each other in concert. Her description of this phenomenon reminds me of a documentary that I saw about the separation of a pair of conjoined twins. The babies were joined along the front of their torsos and, prior to their separation, their mother was the only person who was able to lift and hold them in a way that the babies found comfortable.

However, the mother–infant dance, where the partners in that most vital of relationships can read each other, is not always easy. In fact, Carol says she's often stunned by how often it simply doesn't happen. Much of her role revolves around teaching a mother how to read her baby's signals of distress. This is more difficult than it sounds, given that preterm babies cry much less than babies born at full-term. However, with a little patience and care, parents can be taught to recognise when their baby is stressed and unhappy—dysregulated, as Carol calls it.

These signals may include a screwed-up facial expression, clenched fists, stiff arms, and even the colour of the baby's skin changing so that it becomes mottled, red, or sometimes blue. In the face of these changes in their baby's behaviour and appearance, the mother is encouraged to slow her movements down, pacing her actions and the way she touches the baby to a tempo that it can cope with.

Carol's observations of babies and their mothers are helpful to her now, as a grandmother, and they have informed the way she went about bonding with her grandchildren when they were born. Of course, she says, she was just dying to immediately touch and hold them but, because of her work with prem babies and their mothers, she was able to pace her approach to her grandchildren, reading the subtle signals they gave that let her know she was proceeding at the rate they were comfortable with.

Often, the way she advises parents to interact with their prematurely born baby may actually result in them touching their babies less. She tells me a story about a father she observed who was visiting his little daughter in the hospital while she was still confined to an isolette. 'He was rubbing his thumb up and down the sole of her foot and every time he did it, she moved her foot away. So he'd follow her and he'd do it again, very lovingly, *lovingly,*' she emphasises, 'touching this little girl.'

'Does she like that?' Carol asked the father, hoping to prompt him to really observe his daughter and read the signals she was sending.

'Oh, yeah,' he told her, 'you can tell she likes it, because every time I do it she moves.'

Carol pauses and for a moment we both contemplate the scene she has just described. 'Can you see how,' she asks me, 'with the best will in the world, parents can actually make their babies move away from them rather then towards them?'

Preliminary research that Carol and her team have conducted into the brain development of the preterm babies whose mothers they have been working with has indicated some interesting results. They were hoping to see increased brain volume in the babies but, while this did not appear to be occurring, what they did observe was that the molecules of the myelinated nerve cells in the brain, through which electrical impulses travel, showed an improvement in the way that they lined up, allowing for enhanced transmission of impulses.

Carol is familiar with the work of Tiffany Field. She tells me about a study research that Field did in the area of newborns and massage where babies were massaged by researchers for 15 minutes three times a day for ten days. The treatment also included gently moving the babies' limbs. At the end of the ten-day period, the massaged babies had put on 47 per cent more weight than the control babies who were not massaged. Kangaroo care can achieve similar results, Carol believes. 'I think human beings need close bodies. Whether you're a baby or an old person, you need people who love you to be physically close to you.'

'There's such a lot of similarity between little bodies that are at the edge of viability and old bodies that are at the edge of viability,' Carol tells me, and this belief influenced the way that she interacted with her father when he was old and dying. Towards the end of his life, he was in a nursing home where she would visit him three or four times a week. But, she says, she would never just sit beside the bed. Ignoring the sometimes-puzzled looks of the staff and other visitors, she would climb up on the bed and sit beside her father so that her body was touching his. She would also stroke his arms and head with her hands.

'There's almost some sort of taboo about touching your dad the way you would touch a baby,' she says, 'but I would do that all that time.' It was something they both found comforting.

On one visit, she found him bruised and battered in his bed after a bad fall. As usual, she had arrived in time to help him with his evening meal, but he refused to eat.

'I just leant over and gave him a big hug,' she says, 'and his old arms went around me.' They stayed like that for a long time and then, finally, when they broke the embrace, she was able to get her father to eat. 'He absolutely needed the close contact more than he needed food. He needed someone to hug him and to understand how he was hurting.'

TOUCH HAS ALWAYS been associated with the healing professions. Doctors, physiotherapists, nurses, myotherapists and masseurs, occupational therapists, and practitioners of alternative medicines use touch diagnostically, to comfort, and in the course of treatment.

In many traditional blessings, the hand of the holy person is placed on the head of the one to be blessed, the touch conferring authority, absolution, or a benediction. The Bible contains numerous references to both Jesus and his apostles 'laying on hands' in order to heal or to bestow the Holy Spirit.

Royalty were once thought to have the power to heal through touch. When the king or queen claimed the source of their right to rule as coming directly from God, their power was exemplified through the notion of the Royal Touch. Exercised by monarchs in England and France up to the 18th and 19th centuries respectively, the ritual was deemed particularly beneficial for sufferers of scrofula, a nasty tubercular condition that afflicted the lymph nodes of the neck area. Ugly, weeping sores resulted, and the condition was known in France as '*mal le roi*'—the King's Evil. As well as being able to bestow healing with the touch of their hands, true kings are apparently never attacked by lions.

Massage, whether therapeutic, remedial, or for relaxation, is a popular way that we can get permission to be touched. As our society has become more affluent, there has been a rise in businesses that enable us to be 'professionally' touched in an impersonal and yet intimate way. From the hairdresser and the beauty therapist to masseurs and physiotherapists, we lower the usual boundary of personal space to allow these strangers and acquaintances to put their hands on our bodies.

Recently, when I had a massage as a treat for my birthday,

as the therapist, a young woman who I had never met before, tucked the soft, thick towel into the band of my underpants and edged them down, I was reminded of the intimacies of touch that happen between a child and its mother. The physical environs of the modern massage suite serve to further evoke the nursery: the darkened room, the soothing music, soft cloths placed against the skin, and the sweet smells of citrus and lavender.

It was like being a toddler again, being fussed over and coddled a little. I was positively tucked in. As I lay there happily submitting to the firm, professional stroke and kneading of the masseuse's hands, I reflected on the pleasant selfishness of submitting to this touch that did not require reciprocation. Apart from the fee that I would pay at the end of my session, I had no obligation to respond to the firm strokes and the occasional deep bite of pressure on a particularly knotted muscle. My experience of it was as a one-way transaction, but I did wonder if my masseuse felt the same way.

Anne Davies is a myotherapist. Myotherapists use many traditional massage techniques, along with a detailed knowledge of anatomy and the working of joints and muscles, to treat pain, injury, and dysfunction of movement. Anne's client base comes mainly from those with sports injuries or postural issues and ranges from elite athletes to weekend joggers. Small and compact, she strikes me as someone who focuses on the biomechanics of massage and its therapeutic benefits to the muscles beneath her hands. I imagine her treatment room as light and spare, totally devoid of candlelight and whale song.

For Anne, when assessing injury in her client, touch is way to 'see' into the muscles and joints. The knowledge she gains through palpating the tissue beneath the skin allows her to gauge where there's inflammation or a build up of fluid, or where the muscle is in spasm.

At one point, Anne leans across the table to take my hand, 'Can I touch you?' she asks, and I give permission. She holds my hand lightly in hers, her fingers moving in small circles over the place where the skin lies close to the bones, uncushioned by a layer of fat. The contact is sure, deft, and professional, as she demonstrates the light pressure that is required to encourage lymphatic drainage. At her assured touch, I find myself responding. It is immediate; Anne's touch is at once reassuring and pleasant. I feel like a cat that finds itself unexpectedly, yet pleasantly, scratched behind the ears.

Despite her clinical approach, Anne concedes that there is often an exchange of energy between a masseur and the person being massaged. This exchange can go either way and sometimes, she says, there are clients who emanate a negative or malevolent charge, and 'you have to protect yourself against that'. Other times, she says, before she begins a massage, she will feel tired and depleted, but by the end of the hour she'll be energised. Something has happened in the time she has had her hands on the skin that has revitalised both her and her client.

No matter who it is on the table before her, there is always an undercurrent of a sexual or cultural nature, she says. Teenage boys, prickly and full of juice, are an interesting challenge as clients. Anne tries to overcome any awkwardness brought on by their partial disrobing and her hands on their skin by talking to them 'like an aunt'. If they have an iPod, she asks them about the music they're listening to; if it's a sports injury that they've come to see her about, then she'll talk to them about the sport they play.

For some, she says, she feels like a hairdresser, in that they talk to her about their families, their work, and the other matters that have filled their day. For other clients, with injuries requiring painful manipulation, she imagines they view coming to see her like a visit to the dentist. She does concede, though, that the release of adrenaline and endorphins that a massage can give, even though there may be a degree of pain, can be pleasurable. And at the memory of her fleeting touch on my skin, I don't imagine a massage from Anne would be all bad.

Towards the end of our chat, Anne mentions a 'skin hunger' that she has noticed, mainly in her older clients. Perhaps they have found themselves single after a divorce or the death of a partner, or it maybe they simply don't touch their partner much any more. Often, she finds, they will spontaneously tell her that it feels good to be touched, but more frequently she is simply aware that the feel of hands on their skin is something that they have been craving. An image of Carol sitting beside her elderly father in his bed in the nursing home comes to mind. I see her with her weight comfortably pressed against his body and her hands stroking his arm.

What are the rules for professional touch? In the massage suite, there are the discretions of the averted gaze and the carefully placed towel that shields areas of the unclothed body that are not being massaged.

In a conversation with Jenny Webb, whose career as a masseuse, myotherapist, and a teacher of both, spans almost 20 years, she tells me that she advises her students that any time they touch a client it must be for a professional purpose. For instance, she warns them against leaving one hand 'resting' on the client's body while the other is reaching for a bottle of oil. She, too, finds that the enforced intimacy between masseuse and their client often triggers a confessional atmosphere that sees the one on the table revealing more than perhaps she wants to hear: 'Some people feel really strange lying on a table, getting touched by someone they don't know and not talking to them.'

This situation, two people together in a room, one of them at least partially naked and the other with their hands on their skin, can create an atmosphere of familiarity that is partially misplaced. Jenny maintains a professional distance by not discouraging her client to speak, but by limiting her own responses and engagement with her clients' conversation. Of course, there are other clients, and I am one of these, who feel no compunction to speak, but submit to the professional touch in silence.

Silence or not, Jenny, like Anne, acknowledges that a transference of energy can happen when skin meets skin: 'I know sometimes when I've had a bad day, I've got to stop myself mentally before I go in, not to take that with me. Who knows whether it comes out through the hands or not.'

Another thing Jenny does, in a small ritual of self-protection, is shake out her hands at the end of a massage, so that, metaphorically at least, she is shedding any stress that has jumped from their body to hers through the sensitive, permeable membrane of the skin.

I ask Jenny how she chose massage as her career, and she recalls, as a child, rubbing her mother's shoulders or brushing her hair. Her father was a sportsman, and the healing and therapeutic aspects of touch were embraced by her family. She remembers her grandmother's hands, the skin of her palms and fingers roughened by gardening, drawing circles on her back to put her to sleep. It was a ritual of her childhood, and the contract was 100 strokes of her grandmother's hand as Jenny lay in her bed.

Her eyes close for a moment and her head falls to the side as she remembers the sensation. 'I can still feel it,' she says, 'I want my 100 back rubs every night, please.'

Touch, of course, is not always a positive experience. Violence and sexual assault leave their own reverberations in the victim, and perhaps in the perpetrator.

Over the last few decades, there has been a growing awareness of the extent of sexual abuse of children. Children who have been sexually abused may become confused about the difference between abusive and caring touch, and who would be surprised at this? When touch can be about what someone wants from you, a transaction where you end up with less than nothing; where the feel of someone's skin against your own can rob you of something intangible and make you suspicious of anyone who puts their hand upon your arm, the world must seem a treacherous and uncertain place.

How easy must it be for such a child to come to regard touch as tool to coerce and control, or as a means to achieve closeness, irrespective of the nature of the relationship? If the skin has its own memory, can the experience of such a betrayal of trust ever be sloughed from their skin as time passes, rubbed away by the love of their family and their own resilience, or does it remain embedded in their skin forever?

Melting Pot:
the colour of skin

> 'I have a dream that my four little children will one day live in a nation where they will not be judged by the color of their skin but by the content of their character.'
> —Martin Luther King, 1963

When the 2000 Sydney Olympic Games were on, our family watched a lot of the swimming competition on TV. My eldest son, who was about three years old at the time, was fascinated with the obvious physical differences between the swimmers, and would often ask of any swimmer who wasn't white, 'Where are they from?'

By the time the finals of the various swimming events were reached, almost all the non-Caucasian swimmers were Japanese, and we answered accordingly. Somehow, over the weeks that we watched the Olympics, my son came to assume that anyone who looked racially different from himself was, by default, Japanese.

'Is he Japanese?' he'd ask in his clear, chiming voice, pointing at the tall, young man with ebony skin and black curly hair busily packing supermarket shelves as we wheeled our trolley past. 'Is she Japanese?' again with extended index finger and bell-like tone as a sari-clad woman with honey-coloured skin climbed aboard the tram.

I tried to explain to him that these people, in all likelihood, were Australian. That just because someone had dark skin didn't mean they were foreign; that the original inhabitants of this continent were all black or brown.

'Remember how we watched Cathy Freeman light the flame at the opening of the Olympic Games?' I asked him. 'She's brown and she's Australian.' I explained that our family looked the way we did because our ancestors came from Ireland. 'Australians come in all colours,' I said, appalled that somehow this wasn't obvious to him. After all, we Melburnians pride ourselves on living in one of the most multicultural cities in the world. There were black kids and Asian kids at his childcare centre. How was it that being white had, for him, even at his tender age, become intrinsic to being Australian?

I counted it as a small victory when one day, not long after this conversation, as we were walking around our neighbourhood, we came across a man delivering advertising brochures.

'Look, Mum,' my son said loudly, pointing at him as we passed, 'He's brown and he's Australian.'

WHEN WE REFER to ourselves and others as being 'black' or 'white', we are rarely talking about just the colour of our skin. Blackness and whiteness each have their own set of inherent meanings to do with history, culture, and politics that go beyond the amount of melanin in a particular individual's skin.

Despite the fact that the majority of the world's population come in varying degrees of brown, it is the binary alternatives of black and white, and their attached symbolism, that is forced upon us. Traditionally, black has been associated with sin and defilement, white with goodness and purity.

As a child brought up within the Roman Catholic faith, I knew that the innermost part of my being, my soul, was, of course, white. Visible only to God, it hovered inside the borders of my body, a perfect simulacrum, albeit a pale, insubstantial one. Sins, I was told, would appear like black stains on that pristine whiteness if I strayed from the path defined as right and good by the Church. Limbo was still a tenet of the Catholic faith then, and for my peers and me it was filled with little black babies—the progeny of the pagan hordes who lived in the dark continent of Africa. Having died before they could be baptised, these babies could never be admitted to heaven but were to linger forever in the in-between place of Limbo.

In my heart of hearts, I envied them. We were told by the nuns who gave us religious instruction that Limbo was exactly like heaven, except that those in Limbo were denied seeing the face of God. That would have been a blessed relief as far as I was concerned. The idea of God's huge, bearded head looming perpetually out of the clouds was alarming to my five-year-old self to say the least.

Even recalling that image now, planted so vividly in my imagination as I sat in the cold room beneath the nave receiving religious instruction, makes me nervous about the idea of heaven. Oh, to be a little black baby, babbling happily in the benign environment of Limbo, blissfully ignorant of the existence of God.

AS I BOARD a bus in Strasbourg, France, my eyes are drawn to an extraordinary looking woman who has already taken her seat. Her skin is exceptionally pale, and her thick, wiry hair, arrayed in an intricate pattern of plaited rows, is a yellowy-cream. When she removes her sunglasses, for a moment, I see that long, milky eyelashes frame her whey-coloured eyes in a remarkable frill. Despite the translucence of her skin and corn-silk coloured hair, her features are unmistakeably African.

The disjunction of her complexion and her features draws my gaze. I fight the urge to stare, but my eyes are drawn again and again to that impossibly white skin and improbably coloured hair. The woman has an otherworldly beauty that sets her apart whether I look at her or not. The effect of her appearance is so dramatic that I immediately associate it with the images once so popular with Benetton, the clothing manufacturer, and its iridescent advertising campaigns featuring gorgeous models with every possible variation of skin and hair colour. I realise of course that, far from being a prop for a hip, multiracial, we-are-the-world clothing manufacturer, my fellow passenger has albinism.

In *The Book of Enoch*, a work regarded as non-canonical by the vast majority of Christian churches and not included either in the Jewish scriptures or the Christian Old Testament, the story is told of the birth of Noah. There is a fairytale quality to the account, with the baby described as having a body as 'white as snow and red as the blooming of a rose, and the hair of his head and his long locks were white as wool, and his eyes beautiful'. The birth of the child sent his father, Lamech, into a spin. He assumed from the baby's appearance that the child was an angel. More modern interpretations of the text take a less whimsical approach, assuming the baby's 'whiteness' is an indication that Noah had albinism.

The Reverend William Archibald Spooner, famous (perhaps apocryphally) for proposing a toast to our 'queer old dean' rather than to the royal lady intended (among other malapropisms), also had the condition.

Albinism is a hereditary condition affecting about one in 17,000 people that results from a lack, or a decreased amount, of tyrosinase, an enzyme necessary for the production of the pigment melanin. This lack of pigment means that people born with the condition may have white or very light hair, pale eyes, and very pale skin. While the popular image of someone with albinism is as red-eyed, most have blue eyes — although eye colour in those with albinism can range from pinkish through to violet, and even to hazel or brown. Apart from the effect on the production of pigment, the other main impact is on the optic-fibre pathways, and people with albinism often have poor vision.

In the past, albinism was considered freakish enough for P. T. Barnum to include a family with the condition in his American Museum and in his travelling sideshow, along with bearded ladies and conjoined twins. Folklore and myth regarding albinism include the beliefs that individuals with it can conduct electricity, read minds, and see in the dark.

Less benign misconceptions are that those with albinism are of below-average intelligence and are sterile. In the Dutch language, *kakerlak*, which also means cockroach, is the word for someone with albinism. More poetically, in some Native American cultures, they are called 'Children of the Moon', because of their aversion to strong sunlight.

During the rise of Nazism in Germany, those with albinism were maligned as 'effeminate', and so were despised. More recently, Dan Brown's novel *The Da Vinci Code* drew on the popularly negative connotations of those with the condition in his portrayal of Silas, the murderous albino monk.

In a city like Melbourne, where people have come from across the hemisphere and from every continent, the range of skin colours among us is nothing short of remarkable. As the lunchtime crowds swarm across the streets, a cross-section of the races of man and the medleys of skin colour form a mosaic: from the blue-blackness of someone who hails from the horn of Africa to the pale-skinned, yellow-haired individual whose forbears were Vikings, and all the shades in between. You would need to plunder a paint catalogue to find names to accurately define the variations in tone, despite the fact that, of our 30,000-odd chromosomes, only two of them carry the handful of genes that determine our skin colour. Yet, somehow, the colour of a person's skin, for many of us, brings with it a whole range of associations that have no basis in fact but merely provide convenient shorthand to compartmentalise difference.

Despite all the permutations of skin colour represented in my city, I find my eyes are still drawn irresistibly to those whose skin tone varies dramatically from my own. There is something about that difference that catalyses my interest, my fascination. Or is it simply my innate racism? Difference excites us all at some level, I suppose—a difference in gender, in status, in sexual orientation. But I am disappointed in my own shallowness—that such a superficial difference as skin colour attracts my gaze so unerringly.

Whiteness is my default position: the times I have spent in countries where my particular shade of pale is not the norm have been brief—and cocooned in the protective swathe of tourist dollars.

Not that I think I'm alone. Photographers and clothes stylists appear to dote on the contrast of white and black: think of Robert Mapplethorpe's photographs of nude men, their sculptured bodies appearing like negatives of each other, studies in sameness and difference. His now almost-iconic images continue to exert a strong aesthetic influence, evidenced in fashion spreads in glossy magazines that often feature a Scandinavian-looking beauty in the arms of a coal-black Adonis.

WHEN TRAVELLING IN VIETNAM, I became aware that pale skin is deemed attractive there—especially by young women. In that hot, humid climate, they take great care to cover up their skin to protect it from the darkening effects of the sun. The ubiquitous motorbike is the most popular form of transport, and apart from the sheer number of these machines, and the fact that whole families seem to travel on them, the most notable thing is the way that the young women who ride on them are dressed. Adopting gangster-chic, they pull on long gloves that reach almost to their shoulders; their hats are worn low over their faces, and kerchiefs are tied outlaw-style below their eyes. Our guide in Ho Chi Minh City, a vivacious young woman called Anh, joked that she didn't have a boyfriend because she was 'too brown'.

As well as being a guide for tourists, Anh worked in a local orphanage. She was slightly bemused by the apparent preference that childless Western couples, coming to Vietnam in the hope of adopting a baby, exhibited towards the darker-skinned Khmer babies at the orphanage. She and the other orphanage staff would try to draw the attention of the prospective adoptive parents to the fairer-skinned babies, whom the locals considered more beautiful.

Like the young women I saw in Vietnam, my grandmother's generation would barely venture outside without gloves and extravagantly large hats to protect their skin from the sun. Freckles and tanning were blemishes to be blanched with buttermilk, if you were careless enough to expose your skin to the sun's damaging beams.

When I was a teenager, we offered ourselves up to the same sun with a sensual slavishness that would have appalled my grandmother and her sisters. My friends and I envied those girls with naturally olive skin: our firm belief was that tans made us look thinner and added to our attractiveness just as surely as a perky bosom. Sunburn, blisters, and skin that peeled off like sheets of dried glue didn't deter us from achieving our aim of the perfect tan. We'd sit in a row in the sunniest position during our school lunchbreak, the skirts of our uniforms hitched high, socks pulled low, to get the nut-brown legs we desired.

Now, of course, we know the dangers of melanoma and skin cancer and, while some of us heed the warnings and cover up in the sun, rates of skin cancer are still very high in Australia. Some opt for spray-on tans, bending over in G-strings to submit themselves to the mist of oil and pigment that will give them that all-over burnish even in their crevasses and creases, no matter if the sun's rays would shine there naturally.

IN EUROPEAN CULTURE, the idea of the dusky maiden and the delights that she might offer to a white-skinned paramour was enhanced by tales brought back by sailors on voyages to the Pacific, such as those undertaken by Captain Cook. The perceived promiscuity and lasciviousness of the black woman was the source of many a fantasy for European man.

This idea appears still to have currency, according to an article in the *New Yorker* by Zadie Smith, a British novelist with an English father and a Jamaican mother. Smith wrote, 'If you are brown and decide to date a British man, sooner or later he will present you with a Paul Gauguin.' Not just any Paul Gauguin, of course, but one from his Tahitian period when he was painting brown-skinned women with frangipanis behind their ears, semi-nude, and reclining on beaches or holding large fruits. It might be in the form of a valentine card, Smith says, or wrapping paper, but it will come. Smith went on to describe a holiday with her lover where she completely destroyed his fantasy of being on a tropical island with his very own brown girl. Rather than the sarong-swathed nymph of his imagination, Smith turned into a swollen, whining wet blanket when she found herself to be allergic to 'the whole country' of Tonga.

Perhaps it is a case of the grass always being greener; in some Melanesian cultures, it is women with fairer skin who are favoured by men. They are deemed to be more sexually receptive than their darker-skinned counterparts, and therefore more desirable.

AS PEOPLE AND POPULATIONS become more mobile, it becomes less and less possible to guess someone's nationality by the colour of their skin. How you sound is more reliable in determining your nationality than what you look like. Skin tone may hold a clue to your ancestors' birthplace, but open your mouth and the sounds that issue forth will place you unerringly where you were bred if not born.

An article published in the *Scotsman* newspaper in early 2004 quoted a Scottish national survey that found that the majority of Scots were more likely to accept someone with dark skin and a Scottish accent as being Scottish than they were of accepting someone from England who had settled permanently in Scotland as being Scottish. The oldest enmities obviously die hardest.

From the earliest times, differing skin colour has been viewed as a problem with no obvious solution. When Europeans first began encountering those with darker-hued skin than their own, various theories were put forward to account for humans coming in different colours. Of equal interest to them was what could be inferred about a person from the darkness or paleness of their skin. Of course, there was a vested interest in defining non-Europeans as lesser beings than Europeans. Predictably racist theories, taking the view that the darker-skinned races were less evolved than the white man, gained support and extended to the hypothesis that those races were a different (and inferior) species altogether.

With the Enlightenment and the elevation of science that it brought, these views of the races, and their inherent differences, were supported by 'scientific' theories and evidence that persisted into the 20th century. In 1927, in his pamphlet titled 'The Characters of the Human Skin in Their Relations to Questions of Race and Health', H. J. Fleure, a professor of geography and anthropology at the University College of Wales, acknowledged the common ancestry of the world's peoples, but took great pains to explain the different characteristics of the races. His view was that it was the superiority of the European environment that accounted for the more 'agile brain of the European'. By contrast, the 'specialization of both Mongolian and Negro' skin, limiting 'the multiplication of sensory endings', allowed for their survival in the sweltering climate of the tropics and was responsible for their 'lesser general irritability' and 'greater equanimity of temperament'.

Because Europeans lacked these adaptations of the skin, Fleure lectured on the unsuitability of tropical climates for the European. He suggested the compromise of a 'sheltered life' for those daring or foolhardy enough to venture to the 'torrid' regions and encouraged the 'employment of coloured native field-labour'. Naturally, he warned, the 'social dangers' of such an arrangement should not be underestimated.

How could any reasonable person possibly take offence at these carefully constructed arguments? Such a flurry of pseudo-scientific claims dressed up in 'rational' language carefully avoids any blatantly crude racist rhetoric. Dispiritingly, equally racist views are expressed in less rational terms on the websites of White Pride and similar organisations today.

THE REASONS for different races exhibiting different skin tones is still of interest to scientists and anthropologists. They seek to understand why such variation has developed between various populations in different latitudes of the globe. With poignant and yet naive simplicity, some even preface or append their theories with the hope that, by reducing skin colour to a response to environmental conditions, racism will somehow become a thing of the past.

The colour of human skin is mainly determined by the amount and size of the melanin molecules found within it. Layers of fat and the presence of blood vessels also play a part. Melanin is a pigment produced by cells called melanocytes, which are present in the inner layers of the epidermis. Sunlight stimulates the production of melanin, which is why skin tans when it is exposed to the sun.

Scientists, anthropologists, and casual observers alike have noted that, historically, darker-skinned peoples were found in latitudes close to the equator. An obvious and convenient explanation was that humans living in these latitudes developed dark skin as a protective device against skin cancer and melanoma, both of which may be caused by exposure to ultraviolet, or UV, light found in greater quantities in equatorial regions. While this theory has its merits, as melanin does act as a sunscreen, on its own it doesn't account for those dark-skinned populations who went north gradually losing colour. If the likelihood of developing skin cancer were the only factor in determining concentrations of melanin in a population, it wouldn't necessarily be detrimental to have darker skin in an area with lower levels of UV radiation. Other factors must have been at play.

Recent research into the colour of human skin begins with the assumption that our earliest common ancestors' skin was fair. This assumption is partly based on the fact that our nearest relatives in terms of species, chimpanzees, have light skin. At birth, young chimps have pink skin on their faces, the palms of their hands, and the soles of their feet. As they age, their skin freckles and darkens due to exposure to the sun.

It is widely believed that as our ancestors moved out of the jungles and onto the sun-exposed savannah and began walking and running long distances, the issue of over-heating became more pressing. Gradually, as these creatures evolved to become modern-day humans, they became less hairy, with an increased number of sweat glands to allow them to more easily regulate their body temperature in that harsher climate. Long, thick hair remained on their heads as protection against their brains overheating. However, with the loss of body hair, their skins were much more exposed to the sun and the risks that it brings.

This leads neatly to the theory that our earliest light-skinned ancestors gradually evolved to have dark skin as a protection against sun cancer.

For American anthropologist Nina Jablonski and her husband, George Chaplin, a geographic information-systems specialist, the sun–cancer theory didn't go far enough in explaining variations in skin colour. Jablonski believed that, as most sun cancers occur in people after their childbearing years, the impact on reproductive success, and therefore on evolution, would be minor. This point was crucial in her search to find other factors that would explain the prevalence of dark-skinned populations closer to the equator and fairer-skinned populations in the north.

Since the 1960s, scientists have known that UV light converts cholesterol in the skin into vitamin D. They have also known that melanin inhibits vitamin D production in the skin, and so that lighter skin allows for greater absorption of UV light for vitamin D conversion than darker skin does. Vitamin D is essential for the absorption of calcium, which, as practically every Australian school child could tell you, is essential for strong, healthy bones and teeth. People with calcium deficiency run the risk of developing diseases such as rickets and osteoporosis. Rickets mainly affects the young, as their skeletons are developing, resulting in soft or deformed bones. Osteoporosis, a risk in later years, results in brittle bones.

While vitamin D is available from some foods, including fish oil, most of us get the bulk of our requirements indirectly from sunlight, which carries UV light to our skin. Although rickets is relatively rare in Australia, in 2001, the *Medical Journal of Australia* identified a new high-risk group for vitamin D deficiency: dark-skinned women and women who wear traditional Islamic dress.

These women, because of the role that melanin plays in inhibiting vitamin D production in the skin, and because most clothing absorbs ultraviolet B radiation (necessary for producing vitamin D), are susceptible to this deficiency, and so, too, are their children.

Jablonski was able, through access to a NASA database of measurements of UV radiation at the earth's surface, to establish that the middle latitudes do in fact get more UV light. Here, where historically people have dark skin, there is sufficient UV to synthesise vitamin D all year round. In subtropical and temperate regions, where people with lighter skin but with the ability to tan are historically found, all but one month of the year provided enough sunlight to synthesise vitamin D; and in the third region, near the poles, on average, there is insufficient UV radiation to produce vitamin D in the skin.

From another study, Jablonksi came across a crucial link between successful reproduction and skin colour. This study showed that folate, a substance necessary to prevent neural-tube defects like spina bifida occurring in developing foetuses, is depleted in the human body when it is exposed to sunlight. Folate is also vital for the production of viable sperm. The lighter skinned a person is, the more rapidly folate breaks down within the body.

With this study, Jablonksi had found a link between factors for successful breeding and exposure to the sun. On the exposed plains of the savannah, darker skin would have meant more success at breeding, and so natural selection would have favoured those more swarthy individuals. Jablonski's hypothesis was further strengthened when she came across a paper written by an Argentinean paediatrician who attended the birth of three babies with neural-tube defects whose mothers had all used solariums in the early stages of their pregnancies.

Because of the large amounts of sunlight available all through the year in those central latitudes around the equator, dark-skinned people living in these regions, despite the large amounts of melanin in their skin, were able to absorb enough sunlight for their vitamin D needs, while protecting themselves from folate depletion, sunburn, and skin cancer.

As populations slowly migrated north, the dark-skinned ancestors of what are now the fair-skinned northern Europeans found themselves at risk of developing vitamin D deficiency during the long, dark winters. Not only did they have to cover up more of their skin in order to keep warm, there were also less hours of sunshine in which to absorb UV rays to do the vitamin D conversion. Those with less melanin, and so able to convert more vitamin D, would have had greater reproductive success in the northern climes than those with darker skin. As a result, these populations would have gradually evolved to have lighter coloured skin.

Of course, for every theory there are exceptions, and in this case it is the Inuit people of the Arctic regions and Tibetans. Brown of skin and yet living up near the North Pole—where in mid-winter the sun barely rises above the horizon—how do the Inuit fit into Jablonksi's theory? In evolutionary terms, the Inuit have only been living in the Arctic regions for a relatively short time: 6000 to 10,000 years. They also traditionally have a diet that is rich in vitamin D and calcium—all that fish and seal blubber. So give them another few thousand years and the Inuit will probably be much lighter skinned than they are now.

As for the Tibetans, their skin is lighter than would be predicted by Jablonski's theory. However, they, too, like the Inuit, have lived in their current location for less than 10,000 years and, because of low temperatures experienced on the Tibetan Plateau, they need to wear sufficient clothing to survive. Their lighter skin allows them to absorb more UV light for vitamin D conversion than if their skin was dark.

The process of tanning is one way that humans cope with the competing benefits and risks of exposure to the sun. In summer, darker, tanned skin provides some protection against sunburn and folate depletion; then, in winter, the tan fades to allow for more absorption of sunshine for vitamin D conversion.

So it would appear that the evolution of the different colour of human skin is the result of a complex balancing act to allow for the absorption of UV to produce vitamin D, the maintenance of sufficient amounts of folate to ensure viable sperm and prevent neural-tube defects, and the protection of the skin against cancers. Across all populations, men tend to have darker skin than women of the same racial groups. Jablonski ascribes this to the increased need women have for vitamin D when pregnant and breastfeeding.

But just when you think the whole skin colour thing has been sewn up, along comes another theory. In 2001, an Australian biologist, James Mackintosh, showed in his PhD that melanin acts as an antimicrobial agent in insects. Melanin is quite a 'sticky' molecule, to the extent that it can clog up micro-organisms so that they find it difficult to proliferate. Mackintosh went on to suggest that this may be why, in humans, individuals with dark skin are less likely to suffer from serious skin diseases.

During the Vietnam War, when American soldiers were slogging through the hot, humid region of the Mekong Delta, white-skinned soldiers were three times more likely to develop tropical ulcers than their black-skinned comrades. So the enhanced capacity of black skin to protect against disease might be an additional reason why people in the tropics developed the capacity to produce more melanin. This might also help to explain why large amounts of melanin are found in areas such as the throat, the nasal passages, and the genitals—areas of the body that are rarely exposed to sunlight. So, rather than correlating with latitude or exposure to the sun, dark skin may be linked to temperature and humidity. But what about those theory-bending Inuit? They still seem to be the exception.

Putting all this information together, the colour of our skin probably results from a combination of various evolutionary responses to sunlight, vitamin D, folate, and the antimicrobial properties of melanin. Skin colour has nothing to do with intelligence, integrity, or temperament, but is determined by a handful of genes. So can racism (based on skin colour, at least) finally be relegated to the past as a shameful and erroneous belief that we, as a species, have finally grown out of? Perhaps eventually it will be taken out of our hands altogether. In 2006, *The Age* newspaper reported on the predictions of Dr Oliver Curry of the Darwin@LSE research centre at the London School of Economics, who said that by the year 3006, most humans, due to our increasing interconnectedness, will have a similar brown skin tone.

O, No! It Is an Ever-fix'd Mark: scars, moles, and other blemishes

> And the Lord said to him, 'Therefore, whoever kills Cain, vengeance shall be taken on him sevenfold.' And the Lord set a mark on Cain, lest anyone finding him should kill him.
> — Genesis 4:14–16

TIGHTLY SWADDLED in a sheet, only his head unswathed, it nevertheless took both his father and me to hold him still. Thrashing his little head from side to side, his eyes were wide and staring, straining to take in the meaning of our betrayal. Pain and fear soaked into his face as if it was a sponge. The most fundamental thing that he had learned in his short life—that his parents would protect him from harm—was apparently false. Bewilderingly, we held him down while a stranger repeatedly pierced the skin around his already-throbbing eye with a needle.

The gash at the corner of his left eye was about two centimetres long. The desire for him to have remained safely on my knee rather than toddling across the grimy carpet of the country pub towards his father was a deep physical pull in my chest that left me hunched and aching. When he tripped over his unsteady feet and let out a yell, its outraged tone wasn't sufficiently different from his usual cries, brought on by everyday bumps and knocks, to alarm me. But when I picked him up I could see where the metal leg of the stool had split his soft baby skin in a deep tear from which the blood almost bubbled.

The proprietor, coming in to the bar from the kitchen with our meals in her hands, saw me with my bleeding toddler, and yelled at the bartender to give me a cloth. He looked about him and picked up the sopping rag used to wipe up the spills of beer, and wordlessly held it out towards me. 'A clean cloth!' his boss snapped.

It was obvious the cut needed stitches, despite a well-meaning but slightly sozzled local telling me I was over-reacting; that my not-quite-12-month-old was 'just being a boy', whatever that meant. After several phone calls and conflicting advice, we drove the 70-odd kilometres to the country hospital to be met by the doctor now torturing my son.

A week or so later, our GP took out the stitches, remarking that the doctor who had sutured the cut had done a good job. It was little comfort. My son had been marked for life. He sustained no lasting damage apart from the scar—his eye was fine, there was no infection—but the healed breach marred the fine, unlined smoothness of his skin, and I mourned the perfection it stole.

THE SKIN is a flimsy veil, easily torn and prone to blemishes. It heals imperfectly, and is smattered with spots, scars, freckles, and naevi. Some marks come to us as we grow and age, alighting on our skin like dark birds that will not fly away; others form as we are made, growing with us in the womb.

We all have a scar or two on our bodies, perhaps the result of adolescent acne or a childhood bout of chickenpox. Some scars have less mundane origins, and may be badges of honour or disfiguring cicatrix, remnants of attacks or of battles won with accident or disease. Regardless of its genesis, a scar is a mystery to be solved, a secret to be shared or withheld, a question to be broached when sufficient intimacy has been established. Every scar has a story, and the more dramatic the better. A well-placed scar alludes to past adventures, secret sorrows, or an attitude of derring-do — a fact understood by young male students in Germany during the early-20th century, who proudly displayed the scars that they had received in duels. Some deliberately slashed themselves or aggravated their wounds by pouring wine into them to exacerbate their scarring.

Heroes of romantic novels often have scars that enhance rather than detract from their physical appeal, and are usually an outward sign of a much deeper wound to the psyche.

One of my favourite books from childhood, and one I still re-read as an adult, is Ursula Le Guin's *A Wizard of Earthsea.* The eponymous hero, Ged, bore a terrible scar on his cheek, the result of an encounter with a shadow beast that he summoned from beyond the realms of the dead. At the time that he received it, the scar bore testament to his youthful arrogance and impetuosity; but, as he aged within the story, which extended over a series of books, it marked him as a man who had seen and done much, and gave him an aura of wisdom and mystery. Heady stuff.

Scars are by their very nature inextricably linked to an event: an operation, an accident, an attack, a rite of passage. They mark a specific point in time. They may attest to our courage, our stupidity, our clumsiness, our unhappiness, our lucklessness, or to a fortunate escape.

Scars mark the place where our skin has been breached, laying the way open to infection, loss of blood, or the intrusion of foreign objects into our bodies. They may be constant reminders of our vulnerability, or of events that we might wish we could forget.

Others tell more ordinary tales, but can nonetheless be markers of a particular time or historical circumstance. Like most of my generation, I have circular scar on my upper left arm, about one centimetre in diameter. It is the site of my smallpox vaccination, at one time given to all children in Australia to protect them from the much more disfiguring scars and possible death that resulted from that terrible virus.

ONE OF THE SKIN'S primary roles is to act as a barrier. It helps to protect the body from injury, the loss of fluids and salts, extremes of temperature, toxic substances, and invading bacteria. To fulfil this role effectively, the skin must be extremely efficient at healing itself to ward off even greater injury.

While the main ingredients of skin, the epithelial cells or keratinocytes, excel at healing, the same is not true for some structures within it, including nerve and blood-vessel cells and the hair follicles. Scar tissue does not have all the characteristics of normal skin, being devoid of hair and pigment and without sweat glands.

When the skin or the surface of the internal organs is breached, the body quickly tries to repair the injury by covering the damaged area with a thin layer of tissue called epithelium, which draws the edges of the wound together. In response to the emergency, protein-secreting cells called fibroblasts begin laying down collagen to strengthen the new skin. In the case of a small wound, where the edges of the breach are close together, the scarring will be minimal; but, in a larger wound, the body's emergency-mode response may result in an overgrowth of scar tissue, resulting in a raised, purplish keloid lump. Characteristics of the original wound, including its size, shape, and the cleanliness or otherwise of what caused the injury will also determine the degree of scarring.

In some societies, ritual scarring marks a transition to adulthood, or the achievement of a milestone and the status attached to it. Various tribes of Indigenous Australians, for example, use scarification in initiation rituals, indelibly marking the end of childhood on the body. The ritual may include circumcision and subincision, where the urethra is cut open back towards the scrotum. Scarification may also be done for personal adornment, with cuts on the torso and limbs that are rubbed with ash or clay to increase the resultant scar. Some mourning practices among particular tribal groups also include scarification, making the process of grieving both public and unequivocally physical.

The Nuba of the Sudan elaborately decorate themselves with pigment, oil, ashes, intricate hairstyles, and scarification. Many in the West have become familiar with their imposing physical beauty through images initially made famous by controversial photographer and filmmaker Leni Riefenstahl, who celebrated the fascist aesthetic in her Nazi propaganda films. A strong, youthful body is prized and celebrated within Nuba society, and all adornment aims to expose and celebrate strength, health, and beauty. The stippled scarring on the body of a Nuba woman also signifies her reproductive status and social position.

Other photographers have followed in Riefenstahl's wake, documenting and fetishising the straight forms of these astonishingly physically-impressive people. In one such image, in a glossy-paged book, a Nuba girl stares resolutely ahead, her right hand tensed on the boulder beside her. Blood streams from beneath her breasts from small, even incisions on her torso. She doesn't look at the hand that holds the hook, which is poised to pull her skin up in readiness for the cut, or at the blade prepared to slice.

What is she thinking as her skin is subjected to this succession of small slashes? Is it the knowledge of her enhanced beauty that gives her the will to bear the pain without complaint? Or is the alternative, not to be scarred, so unthinkable that no debate enters her mind? The photograph does not tell us whether she flinches from the ritual, or if she is impatient to undergo it, eager to embrace this sign of her maturity.

At the onset of puberty, a first set of scars is made on a young Nuba woman's torso. She receives a second set, also on her torso, when she begins to menstruate. After the weaning of her first child, she is scarred on her back, arms, buttocks, and the backs of her legs. Both Nuba boys and girls receive scarring on their faces after puberty. Cuts above the eyes are believed to improve eyesight, while cuts on the temples guard against headache.

Scars for the Nuba act as a social code as well as beautification, but for those of us who live in the urban West, scars are more often a testament to trauma, injury, surgery, or suffering.

In contrast to the young women of the Nuba, who proudly display their scars, some young Western women hide the cuts that they inflict upon themselves. These women, who slice their skin in secret and conceal the wounds beneath their clothes, are known as 'cutters'. There's no room for obfuscation in that unsentimental label, a baldly descriptive term for these young women who self-mutilate with razor blades and other sharp implements. The scars mark their alienation from their bodies, and they watch their blood flow as a way of coping with psychological pain.

Young men usually take a different path to act out their turmoil, using drugs, alcohol, or violence against others to kick against the demons that rage within them. Often, girls who harm themselves are survivors of sexual abuse, although this isn't always the case: not all sexual-abuse survivors hurt themselves, and not all cutters have suffered sexual abuse. Still, it is a factor common to many who act out this behaviour.

From the outside, it is hard to understand why someone who has been abused or who is in intense psychological pain would take a blade to their own skin. Why compound the hurt by marking it out in such a drastic way? Why punish yourself for the pain inflicted by someone else?

Wendy, an old friend of mine who was a youth worker for many years, and who wielded the razor on her own body as a young woman, has some insights. She has worked with young women who were survivors of sexual abuse, and says that, while some young women who self-harm may be hoping that someone will notice and intervene, most cutters will hide the scars, taking the blade to areas of their skin that are relatively easy to hide under clothing.

Young women who cut themselves often feel disconnected from their own bodies, Wendy says, and have trouble articulating their feelings. They internalise their hurt, keeping it beneath the skin; they begin to cut when the pain starts to swell within them. It's one way to relieve the pressure and, for some young women, bring a tangible sense of relief and release.

Wendy worked for some time in a refuge for young women

who had suffered sexual abuse. The policy with cutters there, she said, was not to overreact. Usually, they didn't require stitches, and young women who were discovered cutting themselves were expected to dress and bandage their own wounds—not that it was easy to practise cool restraint as these damaged girls punished their own skin for the crimes of others.

Self-harm, like other ways that young women wage war against their bodies, including through eating disorders, can give the illusion of a sort of power. Other aspects of her world may be spiralling out of control, but for those minutes when the razor is in her hand, the cutter is in charge of her body. It's not necessarily a conscious process, Wendy says: 'It's almost like the build up of emotional pain means there has to be an outlet. Otherwise they might go mad or kill themselves.'

Often, these young women are removed from the pain that they are inflicting on themselves, because they are cocooned in the cottonwool of self-protection. They shut part of themselves down, putting a buffer between their experiences and their emotions. Cutting themselves can be an attempt to slice through that numbness, seeking a sensation that is unambiguous, even if that sensation is painful.

'Getting a razor blade and carving into your arm, at least you know you're alive,' says Wendy. 'There's nothing like watching the flow of blood on your skin to see the fact of your life demonstrated.'

It's a view that fits with a psychoanalytic understanding. 'Mutilations of the skin,' according to the French psychoanalyst Didier Anzieu, '... are dramatic attempts to maintain the boundaries of the body and the Ego and to re-establish a sense of being intact and self-cohesive.' It's a paradox, but one that has a certain logic. What better way to demonstrate where your boundary lies than by slicing through it?

And the scar that's left behind? Wendy says it 'reinforces that depth of sadness or potential for madness—a physical reminder of your fucked-up-ness'. This seems too gloomy to me. I think of the Nuba women, the journey of their lives stippled upon their skin like a delicate lace. Mightn't the scars inflicted by their own hands by those troubled young women come to be seen as a mark of their strength, their will to survive, a sign of how far they've come?

Wendy deflates my bubble of optimism. 'Not in my case,' she says. 'I only notice the scar on my wrist when I'm depressed. Still, cutting is not as simple as saying that it's about hating yourself—that's way too simplistic. Perhaps in some states, pain is better than nothing.'

Her view is echoed by another friend who once, early in her career, worked as an assistant psychologist in a forensic hospital for severely disturbed women in the UK. It was not a pleasant experience, and she shudders as she talks about it. At one point, she refers to the inmates as 'prisoners', then corrects herself: 'Patients, I mean.'

Most of them 'had terrible childhoods and they had gone from one awful, horrible, abusive environment to the next until something bad happened. They'd killed people, most of them,' she tells me. The hospital was just the latest in the series of terrible life situations that they had been in. She tells me that she had received death threats, and had nails pushed into her car tyres during the time that she worked there.

'The patients did this?' I ask her.

She rolls her eyes at me. 'No, the nursing staff in the hospital.'

Many of the staff had spent decades working in prisons and were institutionalised to a pathological degree, similar to that of many of the patients. Some of the more brutal abuse had been the subject of an exposé of the institution after someone had smuggled a camera in and captured it on film.

Self-harm was rampant in the hospital, she tells me, and would come in waves. Someone would take a blade to themselves and there would be a rash of copycat slashings. Burning their skin with cigarettes was another method frequently used by the women to act out their frustration, and to alleviate the sheer boredom of incarceration.

My friend's view was that the extreme self-harm that she witnessed was in response not only to the abuse that the women had suffered, and in some instances still suffered, but also to the non-stimulating environment of the hospital. Burning and slashing their skin, and swallowing objects or inserting them under it, was a way of feeling something, of alleviating the terrible boredom and the crushing brutality, of testing and reaffirming their boundaries in the most extreme and drastic of ways. Probably, as for the young cutters mentioned earlier, it was also a means of feeling a semblance of control in an environment where they had none, and a brutal, but effective way of breaking through numbing emotional pain to something that was sharper and almost exquisite.

My friend recalled one patient, a smoker, who had a very large bosom. When the patient sat, she would balance her glass ashtray on its jutting shelf to catch the ash of her cigarette. One day, the patient deliberately smashed the ashtray and inserted the jagged remnant into one of her breasts.

Recently, my friend had heard, they had removed all the

women from that institution and placed them in less-secure environments. One hopes that their boundaries will once again become clear to them, and that they will no longer feel the need to test the edge of their existence by thrusting through their skin with such brutal vehemence.

What happens when the scar is not something that you've chosen? When it represents a trauma and an injury so deep and profound that it speaks for itself, and for the pain that the wearer has suffered? At the trial of Amrozi, who was found guilty of the bombing of the Sari Club and Paddy's Bar in Bali in October 2002, victims of the blast were called as witnesses. In highly emotional accounts, they gave vivid and searing descriptions of what they had seen and experienced that night.

Three Australian survivors of the attacks, which killed 202 people, including 88 Australians, gave heart-rending evidence. One, Peter Hughes, deeply affected the bench when he lifted his shirt, like some of the Balinese victims, to show the extent of his burns.

A newspaper report of the trial described how some of the witnesses chose to remove the pressure bandages from their ravaged bodies to display the injuries and burns they had received in the explosion. This public display of their scars was a silent but potent testament to their suffering. Their scars were not only evidence of their physical injury, but also of their experience of witnessing the terror, the broken bodies, the destruction, and the loss—of sons, daughters, friends, and spouses.

Within Australia, much of the discussion of the Bali bombings was couched in terms of the explosions being an attack on the nation as a whole. Because the attack was seen in the context of terrorism, the scars of the non-Indonesian victims became a physical symbol of an assault against 'the West'. The scars on their damaged bodies are an indelible record not only of the injuries they, as individuals, sustained; they also came to symbolise the assault on the edifice of Western ideals and values.

OUR SKIN is a site of communication, whether we choose it to be or not. A mark made in secret and hidden from the eyes of others will still tell a story even if the only one who sees it is the one who carved the sign.

Much more benign is the scar that we share with all humanity, which is also the first we receive: our navel. Marking the site where the umbilical cord, that conduit for nourishment, was attached, it proclaims that we all have essentially the same beginnings, whatever our growing awareness of differences of culture, class, and wealth.

In some religious circles, however, the origin of the belly button engenders its own peculiar controversy: did Adam have a navel? If Adam was indeed the first man, created from dust, as Genesis tells us, what need would he have had for a belly button? There was no mother's womb from which to draw sustenance, springing as he did from the hand of God.

In the past, artists mindful of the dangers of going against religious orthodoxy avoided the question with the strategic placement of leaves or other convenient props. Legends have evolved to account for Adam developing a belly button after his creation. One of these tells how Adam's navel was formed as God reached into his body to extract the rib from which Eve was made. Of course, this still leaves Eve without a belly button, but this was obviously a less pressing question.

Another story, originating from Turkey, claims that, on seeing the pinnacle of God's creation, Satan spat at Adam, hitting him on the belly with his spittle. Allah moved quickly to remove the polluting substance, tearing out the piece of skin where it landed, and leaving the scar that would become Adam's navel.

A short story that I read many years ago had a group of children wrestling with a similar premise. Some distance down the beach on which they are playing, they see a boy who has merely a beige smoothness in place of the familiar knot of skin that would signify his human birth. Immensely disquieted by his lack, and with increasing consternation, the only conclusion that the children can come to is that he must be an alien. As they move as a group to confront him, they see that a flesh-coloured bandaid is covering his bellybutton, and the mystery is solved. Their relief is enormous.

I have a friend who, though of course she was born with one, came to have her navel removed. As a young woman, she became critically ill, and was hospitalised with an array of baffling symptoms. Unable to come up with a diagnosis, and with growing alarm at her deteriorating condition, the doctors decided to simply cut her open so that they could see what was happening with her internal organs. The necrosed spleen that they removed eventually pointed them in the direction of a diagnosis: lupus. Not a cure, however, because lupus sufferers live with their disease as it progresses, and with the symptoms that become more apparent every year—cerebral haemorrhages, seizures, premature births to name a few. For her, the scar that begins at her sternum and obliterates her belly button as it traverses all the way to her pubis remains an inerasable sign of the invisible disease that she lives with.

IN JUDEO-CHRISTIAN SOCIETIES, marks on the skin have traditionally been seen as somewhat less than godly: at best, as evidence of dubious morality; at worst, as a visible sign of the Devil's influence. In Leviticus (19:28), among the litany of laws given to Moses is one that explicitly forbids the marking of the skin: 'Ye shall not make any cuttings in your flesh for the dead, not print any marks upon you: I am the Lord.'

In Genesis, God set a 'mark' upon Cain, banishing him for the murder of his brother Abel and condemning him to a life as a fugitive. The account of the murder and of Cain's punishment is dealt with a brevity that is both frustrating and tantalising. Few details are given as to the motives of the various players in the story, and there are several apparent contradictions. And what exactly is the 'mark' that God places on Cain in response to his plea that his punishment, and his fear that he will be killed by his enemies, is too great to bear?

Some Biblical scholars propose that Cain may be an ancestor of the Kenites, who used a tattoo to distinguish themselves from other tribes. Others have speculated that the mark may have taken the form of a horn or a facial tic, or perhaps his face was twisted into a scowl.

The purpose of the mark, as well as its form, is debated. Was it so ugly and disfiguring that people avoided him? Did it itch and burn, an ever-present reminder that he was a murderer, guilty of fratricide? Was the mark to protect Cain or was it to humiliate him, to mark him as a fugitive so that strangers would deny him succour? Was it to warn those among whom he wandered that it was his punishment to be a fugitive, and that they were forbidden to kill him? A double-edged sword, the mark would protect him from retribution, but it would also set him apart as one who had sinned.

In one of the contradictions of this passage, despite the mark and his being cursed as a wanderer, Cain married and founded the first city in the Bible, naming it after his son, Enoch.

Somewhat paradoxically, marks on the skin have also been taken to be symbols of divinely bestowed powers, or of affiliations with saints. It was a widespread belief at one time that individuals with particular magical powers or gifts, such as the gift of healing, would have a distinctive mark on their skin, one that they would have carried since birth. Sometimes the mark would take a particular shape that indicated the individual's talent. For example, a mark in the shape of serpent might be found on them if they could heal snakebite. Someone who was born the seventh son of a seventh son was once widely thought to have special healing powers, and might have a cross on his tongue, palate, or thigh. In France, it was thought that seventh sons could be identified by a fleur-de-lis, a heraldic symbol resembling a lily, on their body. Someone of noble birth swapped as a baby may be found to have a mark in the shape of the French royal lily or some other heraldic emblem.

Perhaps in a tradition that originated with the Mark of Cain, the practice of scarring the skin of someone who transgressed dates at least from Biblical times. Such modifications of the skin were seen as a sign of the outcast or the criminal. While in some contemporary urban subcultures branding is emerging as another fetishisation of youthful, unmarked skin, akin to tattooing, in the past it was used as a punishment and to permanently label those regarded as criminals.

In the 16th and 17th centuries in England, a 'v' for vagabond could be burnt into the breast of someone judged able-bodied but who remained idle. As part of their punishment, they became slaves for two years; and if they attempted to escape, they could be branded with an 's' and made slaves for life.

Branding is a sign of ownership, particularly of animals, but also of people. From Mesopotamia to Greece, the West Indies, and North America, slaves were branded so that they could be identified, or as a punishment. Like the Mark of Cain, these brands would scar the unfortunates for life, forever proclaiming their guilt or subservience.

AS A CHILD, sitting on the unforgiving wooden pews of the local Catholic Church every Sunday, I watched for the man with the birthmark. He was young, and the large, ragged patch that covered almost half his face was a cruel disfigurement. It stretched from below his right eye, covering the whole of that side of his face and continuing down his neck, engulfing most of his chin and lower lip. A deep, angry red, it had the appearance of having been plastered onto his face with a spatula, thickening and coarsening his skin so that it looked like a slab of raw meat.

Was God so cruel, I wondered, that he would punish someone by marking him like that? What could he have done to deserve such a thing? His head was always slightly lowered, as if to avoid the sympathetic but always curious gaze of onlookers. It was no use: no matter how he held his head, the birthmark clung to his face and would not be dislodged.

The way a port-wine birthmark like the one I've just described seems to be arbitrarily splashed onto the skin makes it easy to imagine it as an afterthought—something thrown or spilt there, rather than having grown organically. Yet birthmarks are usually congenital, and are caused by an overgrowth of blood vessels. Port-wine stains are permanent, darkening and thickening over time, and no amount of scrubbing will remove them. Some, though, can now be treated successfully with laser therapy or surgery.

Other birthmarks are less disfiguring. The common 'strawberry' naevi bloom and grow on the skin after birth, growing redder and more like the fruit that they are named after. These are rarely permanent, usually gradually reducing and disappearing altogether.

The November 1914 edition of the *Ladies Pictorial* reported the warning of 'a Frenchman' upon seeing the German Emperor Wilhelm II with his withered left arm: 'Distrust those that are marked by the Creator.' His caution echoes an old Scottish proverb that intones: 'Beware of the man whom God hath set a mark on.'

Such admonitions add a whiff of the unholy to any physical blemish. Birthmarks, similarly to the mark that proclaimed Cain's blood-guilt, have traditionally been seen as evidence of sins, fears, or other imaginings: not necessarily of the child, but of its mother. The belief that maternal thoughts could affect the foetus can be traced back to Aristotle and Hippocrates, and survived into the 18th century in folklore: a woman whose child was born particularly hairy might have been frightened by a bear; a child with severe deformities might be the result of its mother dreaming of monsters; or a woman with a child afflicted with a severe birthmark might suffer the disapproving glances of those who assumed that she had committed a grievous sin during her confinement.

Such theories aren't confined to the Middle Ages. A dermatologist told me that when she was pregnant, a friend of hers took it upon herself to bring her potato chips every week.

Her friend had seen the dermatologist eating potato chips and assumed that she had a craving for them. Her friend told her that if a woman craves something during pregnancy and resists eating it, the baby will have a birthmark in the shape of the craved-for food.

Not entirely dissimilar to the theory that maternal impressions can affect the foetus as it grows is the view that past lives can leave impressions on the body of later incarnations.

The late Doctor Ian Stevenson was a physician and psychiatrist, and the founder of the Division of Perpetual Studies at the University of Virginia. In a scenario that would not be out of place in the cult television series *The X-Files*, the Division of Perpetual Studies, a unit within the university's Department of Psychiatric Medicine, investigates apparently paranormal phenomena using scientific methods. Dr Stevenson did extensive research into children who remembered past lives. A number of these children, whom Stevenson interviewed, display birthmarks that correspond to injuries that they describe having suffered in an earlier life.

Stevenson published many articles and several books on his research, always in language tempered by a cautious and measured tone. Stevenson appears to have been meticulous in his research, searching out and interviewing the children concerned, and those that knew them both in their present and supposed previous lives.

One such case involved a boy in India who had been murdered when he was six years old. During the attack, his head was severed from his body. Some years later, his father heard of a boy who had been born six months after his son had died, and who related stories of his earlier life and details of his murder which corresponded with the events that befell his son. The boy also recalled the name of his father in his previous life: Jageshwar.

The boy had a long, thin birthmark on his neck, resembling a scar made by a knife wound. Cue spooky music.

THE CHILL of the wind slips under the door, causing her skin to tighten and pucker. The poor, ragged cloth of her only dress slumps in a grubby heap at her feet. Her world is reduced to the stained weft and warp of the coarse material as she focuses her gaze on it, shutting out the preparations of the stranger behind her. The hushed chatter of the others in the room — the village priest, her mistress, and Samuel the tavern owner, her accuser — suddenly ceases. A sharp, stabbing pain takes her by surprise, making her shriek and jerk away.

'I have no wish to bind you,' she hears the well-bred voice of the stranger say, 'but I will not hesitate if you do not submit to the pricking.'

She does not look at him, but holds her head in her hands, leaning into the rough wall in front of her. Again comes the stabbing pain, and again. The warmth of her blood trickling down her skin is curiously soothing as she bites down on her lip, straining not to cry out.

She tries to follow the flashes of improbable colour that dance and flicker in the darkness as her hands press harder and harder into her eyes. Perhaps she can follow these phantasms of light out of this cold, draughty room, and away from those who would gladly see her burn.

She hears the clatter as the witch-pricker drops his bodkin, but does not see him deftly pick up another at his feet, concealing the first in the folds of his cloak. She only knows that the next thrust is not painful at all, as the metal shaft retracts into the cunningly designed handle. A sharp intake of breath from the others in the room distracts her from her bewilderment.

'She does not bleed, you see, she does not bleed.'

During the mad years of the witch-hunts in the 16th and 17th century, perhaps hundreds of thousands of women were murdered for consorting with Satan. It was held that a witch could be identified by the mark of the Devil on her skin. As the definition of such marks was very imprecise, birthmarks, moles, or other blemishes could be declared marks of the Devil. Some witch-hunters even claimed that it was possible for the marks to be invisible, and able to be found only by pricking the woman's skin all over. If any spot failed to bleed, it was plainly the site of the Devil's mark, and the woman could be declared a witch. It was not unheard of for some witch-prickers to use a device with a retractable blade, alternating it with a real spike.

Other growths, such as an extra nipple or a wart, could also be used to identify a witch. Demons and imps that craved the taste of human blood were said to suckle there. Of course, warts are largely dead skin and will not bleed when pricked—this, and the implication that such growths provide no sustenance to humans but are for the Devil and his evil sprites to feed on, gives rise to the saying, 'Cold as a witch's tit.'

The Mark of the Devil is sometimes extended to include the Mark of the Beast, referred to in the Book of Revelation (13:18): 'Here is wisdom. Let him that hath understanding count the number of the beast: for it is the number of a man; and his number is Six hundred threescore and six.'

These verses, in conjunction with the earlier ones that speak of a 'mark' placed on the right hand or on the forehead, have fuelled urban myths and conspiracy theories based on distrust of the digital world. Commonly, these myths cite supposed plans to implant computer chips in each person on the planet that will record all that person's data, and will also be stored on a giant computer somewhere.

A reference to trade—'no man might buy or sell, save he that had the mark'—adds to the impression that the beast will enslave humanity with our insatiable consumerism.

Various internet sites warn Christians against accepting anything that might be the 'mark of the beast', including devices that store information such as Smart cards or credit cards. According to the *Oxford Companion to the Bible* (*OCB*), popular perceptions of the 'beast' being either the devil or relating to the digital age are false. David H. van Daalen, writing in the *OCB*, names the 'beast' as the personification of the rule of the Roman Emperor Nero. Apparently, when Nero's Greek name is written in Hebrew letters, which also function as numbers, they add up to the diabolical number 666.

MARILYN MONROE had one above the left corner of her mouth. Fashionable European ladies in the late-17th and early-18th century hid their blemishes with fake ones. According to folklore, their appearance on various sites on the body indicates everything from fertility to honour and riches. They are, of course, beauty spots—or, more technically, moles or naevi.

Beauty spots refer to small, dark marks usually found on the face. A beauty spot is considered particularly beguiling when it is situated near facial features that are already associated with beauty: at the corner of the eye or the mouth, or perched on a high, curving cheekbone, a beauty spot can accentuate and define facial allure. Like the dot that completes the downward stroke of an exclamation mark, a beauty spot magnifies the effect of fluttering eyelashes, a sensuous mouth, or the dewy smoothness of young skin.

There is a point, however, where a beauty spot becomes a mole. Too large, too hairy, too bulbous, and they are deemed to detract from, rather than enhance, one's beauty. The witches and wicked stepmothers of fairytales often feature moles, the end of the nose or the chin being favoured spots.

It's not always apparent where this point between beauty mark and mole lies, however. In the early stages of her career, Cindy Crawford, one of the 'supermodels' of the early 1990s, famously rejected advice to have her mole removed, and turned it into her trademark.

In the late-17th century and into the 18th, women of the English and French upper classes embraced the allure of the beauty spot by wearing fake face patches fashioned out of small pieces of gummed black taffeta. Men were not impervious to the advantages of using them, either. Fashioned into pleasing shapes, they could be placed on the face to enhance a beguiling look or, more practically, to cover small-pox scars or other blemishes.

At their height, extremely elaborate shapes developed: coaches with six horses, a pair of lovebirds in a tree, as well as the more muted stars and crescents. In London, they also became a way of declaring one's political inclinations: when a patch was worn on the right cheek, a lady's sympathies could be presumed to be with the Whigs, while on the left, with the Tories.

The *Spectator*, in 1711, reported that the trend was so strong that one particular lady, in drawing up her marriage contract, insisted that, whatever her husband's opinions, 'she shall be at liberty to patch on which side she pleases'.

In some European folkloric traditions, moles, depending on where they are located on the body, may be omens of good luck or indications of a person's character. A mole on the right arm or shoulder is a sign of wisdom, but on the left it indicates someone who is fond of an argument. A mole on the foot or ankle is evidence of courage in a woman and modesty in a man. If you have a mole on the right side of your forehead, or on your right ear, you can expect great riches; but prepare for the poor house if you have mole on the left of these same features.

Moles, particularly in Australia where the incidence of skin cancer is high, are also associated with disease. Formed by an accumulation of melanocytes, or pigment cells, moles, like birthmarks, are usually congenital, but they can also form within a few years after birth. Generally, these lesions are harmless, although those of us with fair skin are especially likely to distrust them, anxiously observing them for any sign that these innocuous clusters of pigment cells have turned against us.

An increase in size or the darkening of a mole may be an indication of malignant melanoma. While only making up a small percentage of skin cancers, melanoma is the most likely to spread to other regions of the body.

In the Museum of Pathology at the University of Melbourne, I have seen melanoma hanging in jars of clear solution, cut from the skin where they sprouted. They are ominous looking things: blackish growths that threaten death. In one glass display vessel, a cross section of a brain shows a secondary, malignant melanoma that had its origins from a primary growth on the face.

I look in the mirror, my fingertips tracing over my own marks: the scars, moles, and assorted naevi that I have accumulated thus far. Even without the benefit of the mirror, I can put a finger on the exact spot where they lie: the large, red naevi that bloomed in the hollow of my left cheek during my first pregnancy, and remained; the scar on the tip of my third finger on my right hand, legacy of reaching into the bathroom cabinet as a child only to connect with my father's razor; the scar on my left breast, left by the incision to remove a benign lump; not to forget the lines on my face that deepen every year as result of the growing laxity of my skin.

There are none that I would part with, despite the mundaneness of their genesis—no tales of heroism or epic struggle do they tell but, nonetheless, they are mine.

Peculiar to Humanity: blushing and tickling

[Blushing] is the colour of virtue.
— Diogenes the Cynic

THE TENET that humans are fundamentally, even intrinsically, different to other animals, is largely unchallenged. However, the precise quality that defines that difference remains somewhat nebulous, and the source of debate. Scientists, anthropologists, sociologists, and clerics have put forward various theories to account for our perceived singularity among the species.

Inherent in this notion of uniqueness is the idea that humans are superior to other animals. For the religious, who have faith in the idea that we are made in God's image, it is our souls that elevate us. Other specialists have cited our intelligence, our highly developed language and cultural systems, and the once-almost-universally-held view that it was our species' ability to use tools that was the hallmark of our humanity.

This last piece of received wisdom has more or less had to be abandoned as an increasing number of animals in the wild and captivity have been observed manipulating objects to achieve a desired outcome. Chimpanzees probe for ants with grass stalks;

buzzard kites crack open eggs with rocks held in their beaks; and female dolphins in Shark Bay teach their daughters to shield their sensitive noses with sea sponges as they forage on the ocean floor.

RATHER THAN THE FACT that we can swat a fly with a rolled-up newspaper, that we are moved to sing and dance, or that we have a spiritual facet of our beings, perhaps it is our very skin that holds the key to our collective personhood, and marks us as subtly, yet unmistakeably, distinct from the scaly, furry, and feathered creatures that we share the planet with.

I'm not referring to the fact that most mammals are much hairier than we are—hippos, those slightly repulsive naked mole-rats that live underground, and whales have even less hair than humans do. No, what it is that makes us different from other animals is that our skin flushes with blood in response to an emotional stimulus, and that we laugh out loud when we are tickled.

The sympathetic nervous system, which is part of our autonomic nervous system, controls, among other things, the smooth muscle of the walls of our blood vessels. This part of the nervous system operates involuntarily. In response to an emotional stimulus, the brain releases chemicals that cause the walls of our blood vessels to dilate, allowing a rush of blood to the capillaries. Sensations of heat and prickling accompany the crimsoning of our skin, which we refer to as a blush.

Blush. It is a round, juicy word that fills the mouth like a piece of fruit. The lips, tongue, teeth, and jaw all must be engaged to pronounce it. From the pout of the plosive opening 'b' to the voluptuous curl and fall of the tongue required for the lateral 'l' and on to the movement of jaw, palette, and tongue that allows for the sibilant finish. That final 'sh', with the sensation of air hissing out between the teeth, lingers in the mouth like the taste of something sweet.

Embedded in the word blush are connotations of lushness and arousal. The warm, delicate reddening of the skin of a ripe peach is described as a blush, and the very word peach is sometimes used as an appreciative epithet for a young girl. The young are more prone to blushing and, as a result, it has associations of innocence and modesty, especially in young women. The sight of a young girl's face suffusing with colour as the result of masculine attention or confusion at a ribald comment was once seen as a pretty indication of young woman's naiveté. In the early-19th century, poet Letitia Landon wrote:

> As beautiful as a woman's blush
> As evanescent too.

Young men, particularly in Victorian novels, tend to 'heighten in colour' rather than 'blush', and then usually in response to an excess of a strong emotion, rather than confusion or virginal embarrassment. Such a description has allusions of vigour and strength; a young man's virility and passion displayed in the ruddiness of his complexion.

It was the naturalist Charles Darwin who recognised the singularity of blushing to our species, observing in his book *Expressions of the Emotions in Man and Animals*: 'Blushing is the most peculiar and the most human of all expressions'.

In *Following the Equator*, Mark Twain expressed it differently: 'Man is the only animal who blushes. Or needs to'.

Even in our current, more promiscuous era, where pornography provides the dominant aesthetic in fashion and music videos, the idea of the 'blushing bride' continues to hold some currency.

For the modern bride, however, pink cheeks might be more accurately ascribed to the two glasses of champagne that she had while getting her hair done. Presumably, in more demure times, it was the contemplation of the bridal bed and the lustful delights that awaited her that brought an attractive glow to the traditional bride's face. Now, most brides are too well acquainted with the gamut of sexual acts for the anticipation of the whole-hearted embrace of their new husband to cause them to blush.

For centuries, women of all ages have also used blusher or rouge to give the appearance of a youthful, blooming complexion. However, blushing hasn't retained quite the romance that is contained in Landon's verse. Flaming red cheeks are now more likely to be associated with shame, humiliation, or embarrassment.

Darwin noted in *Expressions of the Emotions in Man and Animals* that blushing is not caused by a physical prompt. In order to blush, it is our mind and our emotions that must be engaged. More disconcertingly, not only is blushing involuntary, it actually increases the more we wish for it to cease. Blushing deepens the embarrassment that caused it in the first place. We can even blush when we are by ourselves, remembering an embarrassing or shameful moment.

Although blushing is more common in the young, Darwin observed that the very young do not blush. They have yet to learn to care what other people think of them, which Darwin insisted was a necessary ingredient to blushing. It is not simply that we have told a lie that will cause us to blush, but the fact that we are aware that we have been caught in a lie.

When my eldest son was about six years old, I noticed that he had learned to blush while his brother, two years younger, was still free of it. The older boy had learned shame. Even now when I see him blush in response to a reprimand, I know that my castigation has hit the mark. But does his blush signify that he knows what he has done is wrong, or merely that I have succeeded in humiliating him? His blushes make me realise the depth of his emotions more powerfully than his tears, his yells, or his sulking. While I may congratulate myself on having successfully disciplined him, a little part of me dies for having shamed him.

As we grow older, our blushing generally diminishes. Our composure and confidence grow, and we are thrown less into confusion. Handsome rogues aren't as likely to chuck us under our chins and tell us how pretty we are as we grow into maturity and accumulate knowledge, experience, and wrinkles.

Nevertheless, there are adults who continue to blush furiously well into maturity. As callous teenagers, my friends and I would judge a battle against our boarding-school dormitory mistress won once the telltale blotches of a blush appeared. For this unfortunate creature, blushing manifested as ragged spots of red that bloomed on her face and neck while the skin between them paled in comparison. We regarded these blotches as unbidden capitulation signals, and laughed heartlessly to see them.

Less cruelly, a friend in her forties in the heady throes of a new love affair coyly remarked to me, 'I think I'm learning how to blush again.'

IN VICTORIAN TIMES, the appearance of a blush signalled a person's open and honest nature—a capacity for duplicity is seriously undermined by a tendency to blush.

Around the 18th and 19th centuries, when Europeans were

colonising and exploring the Americas, questions of the humanity of the people who lived in these new worlds, and the hierarchy of the races, were being raised and discussed. Physiognomy (the study of the face and its features in determining a person's character) and phrenology (the theory that the shape of your skull indicated your intelligence) were accepted as sciences. In that climate, whether the dark-skinned races could blush was seen as one of the indicators of their affinity to the fairer-skinned Europeans.

Alexander von Humboldt, who recorded his journeys through the 'equinoctial regions' of the Americas in the late-18th and early-19th century in his *Personal Narratives*, asserts that the facial features of Europeans were more mobile as a result of 'the sensibility' of their souls:

> It is only in white men, that the instantaneous penetration of the dermoidal system by the blood can take place; that slight change of the colour of the skin, which adds so powerful an expression of the soul. 'How can those be trusted, who know not how to blush?' says the European, in his inveterate hatred to the Negro and the Indian.

In spite of such attitudes, which held currency during his lifetime, Charles Darwin concluded that all human races blushed.

In contemporary society, we tend to value composure over powerful 'expressions of the soul'. Humiliation and shame, after all, are neither easy nor pleasant to deal with. Self-possession and confidence, rather than a blatant display of one's emotional state, are regarded as essential prerequisites for success. In such a cultural climate, those who blush furiously at the slightest provocation may feel at a disadvantage.

IT'S 9.30 on a Monday morning in the meeting room of a small publishing house. The staff file in, exchanging rundowns of their weekends and clutching their morning coffees. They take their seats, slapping folders and sheets of paper down in front of them.

'Right,' says their boss, 'how are we travelling?'

They go around the table giving updates on production schedules and flagging any anticipated hold-ups.

Stella slips into the chair on her boss's right; this is her preferred position. Her boss consistently gets the person directly on his left to begin the weekly update. Sitting on his right, Stella gets to speak last, and she wants to stave off that moment for as long as she can.

In hectic weeks, if timelines are falling behind, people sometimes leave early in order to get on with their work. Stella prays for unreliable writers, printing delays, and any other disaster that will make at least some of her co-workers excuse themselves before she must speak. This week, however, is a good one as far as deadlines go: clients are reasonable, timelines are being met, no one is leaving early. Now, there are only three people left to give updates before it is her turn to speak.

Stella swallows hard: her respiration rate is increasing, her heart starts to beat faster as her palms become sweaty, and a yawning hole opens in the pit of her stomach. Worse, however, is to come. She knows that the moment she opens her mouth, crimson blotches will bloom on her neck, the skin of her face becoming hot and prickly. Everyone will stare, the telltale reddening of her face broadcasting her incompetence and lack of confidence to her workmates, her boss, and to herself.

She lurches to her feet, mumbles something about feeling sick, and flees the room.

Stella may be fictional, but there are many adults whose propensity for blushing, and for the feelings of humiliation that it engenders, impinges upon their lives to the extent that they seek professional help. For such people, seemingly minor or inconsequential occurrences cause them to blush, and blushing deepens the embarrassment that they feel, paralysing them and making minor events an ordeal. Unexpectedly seeing a neighbour at the supermarket or having to sign a cheque in public can become occasions of stress and anxiety. Rather than put herself in such situations, someone like Stella may sabotage her career or sacrifice her social life simply to avoid blushing.

Severe facial blushing is referred to in medical terminology as Idiopathic Craniofacial Erythema (ICE). Translated into plain English, the term refers to a reddening of the face due to an unknown cause.

Once pathologised, a condition calls for a cure—even if the condition is an ordinary human response rather than a disease. Developing a cure for blushing may seem akin to searching for a cure for laughter or crying. However, when blushing reaches a stage that it prevents people from engaging in everyday situations or participating in activities as they wish to, then some medical practitioners believe their experience lies outside of what can be considered normal. And yes, there is a cure.

A surgical procedure known as Endoscopic Thoracic Sympathectomy (ETS) was first performed to provide relief for those who suffered from excessive sweating of the face, scalp, and hands. Some patients who had undergone the procedure reported that they also no longer blushed. Surgeons were then able to offer the operation to people who suffered from ICE. The procedure, performed under a general anaesthetic, involves making an incision below the armpit that is so small it does not require stitching. Ultra-thin surgical instruments, including a telescope, are then inserted into the patient through this incision. Both lungs are partially deflated to allow identification of the nerve that controls the size of the blood vessels, and the nerve is then clamped with a titanium clip. After the procedure, the lungs are re-inflated.

As with all surgical procedures, there are risks and side effects associated with ETS. Compensatory sweating, where other parts of the body produce more sweat, is the most common side effect of this surgery, even when it is performed to reduce blushing.

Treating blushing as a disease is controversial even within the medical community, and there are means other than surgery of treating severe cases. A vascular physician may prescribe drugs such as those used for high blood pressure to enable people to deal with situations in which they anticipate blushing. For others, mild antidepressants or migraine medication may have the same result.

Dermatologists, too, see patients whose facial blushing causes them distress, including some with diseases that have a flushing component. Rosacea, for example, is characterised by a facial rash, which in most severe cases is accompanied by papules (red spots) and pustules. Those of Celtic origin are more prone to the disease and, as one dermatologist told me, it is all the good things in life that exacerbate it: alcohol, spicy food, coffee, chocolate, and exposure to the sun.

Those with the condition may also develop an enlarged, misshapen nose with the appearance of orange peel, due to enlarged pores and fibrous thickening—a condition called rhinophyma. Those with rosacea and rhinophyma are often made uncomfortable with the knowledge that people frequently assume they are alcoholics. The condition is treatable with antibiotics, and by avoiding vasodilators like alcohol.

Another avenue for those who believe that their blushing is having a negative impact on their lives to an unacceptable degree is to seek help from a psychologist. Generally, in the case of someone like our fictional Stella, blushing would be treated as part of a wider problem of social anxiety. A psychologist, rather than attempting to 'cure' blushing, might seek to help their client understand why they blush, and also to challenge the assumptions that they have about blushing. Success might be judged in terms of the person blushing less often or more mildly, and attaching less importance to the occasional flaming face.

For someone like Stella, who is socially paralysed by this unbidden physical response to an emotional stimulus, blushing becomes a catastrophe: a sign of weakness and incompetence. It has taken on, in her mind, far greater significance than it has for anyone who might observe her blush. By assisting her to be less concerned about gaining the approval of others, and encouraging her to practise taking on situations that she would usually avoid, ideally, her fear of blushing and the circumstances in which it occurs would diminish. If she fears blushing less, she will do it less. Techniques to encourage physical relaxation may also help to relieve emotional anxiety and make her less prone to blushing.

IN *The Expression of the Emotions in Man and Animals*, Darwin quotes a Dr Burgess, who asserts that the Creator designed blushing 'in order that the soul might have sovereign power of displaying in the cheeks the various internal emotions of the moral being'.

Like Alexander von Humboldt, Dr Burgess viewed blushing as a mechanism for the soul to reveal its true emotions on the exterior. It is an idea not without appeal. While I am not so enamoured of the racism imbedded in von Humboldt's views, I have to confess a certain attraction to the notion that the sensibility of our souls should be displayed upon our faces.

For my part, I'd like to see a return to the well-deserved blush. Not so much for sweet, young things prettily pinking up at the gentle teasing of a virile man, but for the older and more powerful in our society. Perhaps if every evasion and attempt at stonewalling was accompanied by a rush of blood to the face our politicians would feel more compelled to reveal their true motivations, and would work harder to earn our trust. Not to mention those shop assistants who avow, 'Oh, they look great on you!' despite all evidence to the contrary.

So, if you are a great blusher, wear it as a badge of honour, a sign of your open and honest nature. And don't forget, a bit of colour in the cheeks can be very attractive.

I SIT IN THE CHAIR at the hairdressers. My new haircut, shorter than my usual style, is almost complete—only the finishing touches to be done.

The hairdresser leans past me to pick up the battery-powered shears lying on the small shelf beneath the mirror in which my haircut and I are reflected. I experience a small frisson of anticipation as the shears vibrate into life with a low hum. I drop my head forward, the back of my neck long and exposed, the hairs on my nape standing up in expectation. A question from her co-worker causes the hairdresser to switch off the shears, and they consult briefly about the location of some hair-care product.

Disappointment floods through me as the anticipated contact fails to ensue. Without warning, the warm hum restarts and, as the cold steel of the blade presses on my skin, the pressure, light yet firm, sweeps up my neck. The delicious shivery tickle that travels up my back is known as knismesis.

'TICKLE ME with your blow,' the child demands, bearing the nape of his neck so that his mother can brush his skin with her lips, lightly feathering his neck with her breath. Almost immediately, he tips his head back and hunches his shoulders, giggling and yelling 'Stop!' only to insist that she do it again.

The mother makes claws of her hands that threaten to dig into the child's sides. At the mere thought of the contact, the child begins to squirm and giggle. As his mother presses her fingertips into the child's ribcage, his laughter intensifies and he draws up his legs and hugs his arms to his sides while wriggling in an attempt to escape. His mother continues to tickle him until he abruptly says, 'Stop, I don't like it.'

Gargalesis: tickling that induces involuntary laughter and squirming.

Being ticklish, like blushing, is one of the characteristics of being a human—or, at least, of being a primate. It's an odd characteristic to consider as being definitive of the uniqueness of humans, and yet one that has been recognised as such for centuries. Shylock the Jew in Shakespeare's *The Merchant of Venice* appeals for recognition of his humanity, no less than that of a Christian, when he asks: 'If you tickle us, do we not laugh?'

Gargalesis is caused by relatively intense pressure repeatedly applied to 'ticklish' areas of the body, like the ribs and the soles of the feet. Few of us would have escaped childhood without being subject to it at some point.

All mammals will show a similar response to a light tickle (knismesis), where the skin becomes slightly itchy or mildly shivering; for example, the way a horse's hide will twitch when the animal is being pestered by flies. It is generally accepted that this response developed to alert animals to the potentially dangerous presence of insects and other hazards on the skin. However, it is only primates that laugh or utter a sound corresponding to a laugh (apes produce a reiterated panting) in response to gargalesis or heavy tickling.

Tickling, like all forms of touch between people, has strict although unspoken rules. An essential element of pleasurable tickling is trust. Tickling is done between parents and their children, among siblings, and between lovers. Generally seen as a non-threatening type of physical touch, tickling is often engaged in as part of the preliminary stages of courtship. Intimates, not casual acquaintances, tickle each other.

Yet, despite the laughter that leaps unbidden from your mouth when you are tickled, it is not always an enjoyable sensation. Tickling provokes a complex reflex response that combines defensive actions with facial expressions that positively reinforce the tickler's actions. In other words, you might be laughing, but at the same time you're protecting your ticklish areas and trying to get away. You may even attempt to kick or strike out at the person tickling you. Children will sometimes demand to be tickled and then run away, only to return to ask to be tickled again. This is their way of controlling the intensity and duration of the tickle.

Being tickled by a stranger, or by someone whom you dislike or fear, engenders feelings of violation, of a boundary being crossed, or even of panic. Children sometimes tickle each other as a torment, and there is evidence that in medieval times tickling, as a form of torture, was used in deadly earnest. A woodcutting from the 17th century depicts a farmer bound by ransacking mercenaries, and left to be tormented by a goat that is licking salt off the soles of his feet.

Few adults, in my experience, claim to enjoy being tickled, although there are exceptions. Some will admit that it depends on who is doing the tickling, and in what context the tickling is taking place. Certainly, there is a distinct but shifting line between the shiver of delight caused by a finger lightly drawn up the underside of an arm and the unpleasant jolt unleashed when fingers suddenly press into your ribs.

Tickling can also be erotic, and a playful component of sexual intimacy. A light tickle might progress to being mildly irritating and then, at some indefinable point, become arousing. Yet I have also spoken to people who simply detest tickling in all its forms, regardless of how, when, or where the tickling is taking place. Certainly, prolonged tickling is generally unpleasant; brevity is desirable, for those of you who were too afraid to ask for tips on successful tickling.

One friend whom I spoke to had memories of being tickled unmercifully as a child. She has a complete anathema to tickling, and is resolved never to inflict it on her own children. Our conversation obviously stirred deeply felt emotions, and the strength of her feelings took me a little aback. Her experience of tickling, she said, bordered on abuse.

TRACE YOUR FINGERTIPS around the soft hollow of your elbow and down your forearm. The skin here is soft and sensitive, largely protected from the sun, and unroughened by the contact that desensitises more exposed areas of the body. As your fingers gently stroke the skin, you will most likely feel a light, pleasant, almost-itching sensation. Perhaps you may even succeed in causing the tiny erectile muscles around your hair follicles to tighten, so that you achieve a goose-bump effect on your arm.

Now lift your arm so that your armpit is exposed, with the fingers of your opposite hand repeatedly pressing into this area, while scraping your fingertips back and forth across it. Any reaction? Probably not. While it is possible to lightly tickle yourself, it is almost impossible to do the same with the laughter-inducing gargalesis. Why is this? Why can't we tickle ourselves? Apparently, when self-tickling, the cerebellum is able to warn that area of the brain that registers touch (the somatosensory cortex) that the body is to be tickled. This knowledge dulls the response to the actual tickling.

It is argued that this aspect of tickling—that it cannot be self-inflicted—gives weight to the theory that it evolved as a tool for social bonding. There is little to be gained from social bonding with yourself. Interestingly, some people with schizophrenia are able to sidestep their brain's machinations, and can tickle themselves.

Other theories, apart from tickling's role in social bonding, exist to account for primates' reaction of laughter combined with an effort to escape. One theory proposes that laughing in response to tickling may be a conditioned response similar to Pavlov's dog salivating at the sound of the bell that it had learned to associate with food. In this scenario, children have learned that tickling takes place in (hopefully) playful situations, and that laughter is appropriate.

Another theory proposes that people are ticklish in those areas of the body that are prone to damage in hand-to-hand combat, and that tickling helps us to develop defensive movements. Christine Harris, an American psychologist with an interest in tickling, posited that the response to tickling might have developed in primates as a way of developing combat skills through physical play.

This last hypothesis is the one that I find the most attractive, purely from an aesthetic point of view. Rehearsing to deflect the thrust of a dagger up and under the arm, or to stop the sinister point of a spear sliding neatly and with deadly precision between the ribs, has a certain romantic appeal. It also has a certain logical validity: a knife slicing through the tendons at the back of the knee—a very ticklish spot for most people—would be more than enough to immobilise a person. Okay, the theory falls down a bit when you consider the soles of the feet, an extremely ticklish area certainly, but not the most obvious site in danger of direct damage in 'hand-to-hand' combat.

Tickling was part of the physical play that we participated in as a family. Along with my brother and sisters, I loved the rough-and-tumble play that our father engaged in with us. My father was a strong man, and we could never get the upper hand through strength even when we ganged up on him, so we would try to tickle him in an attempt to turn the tables. Maddeningly, he was not ticklish.

It's a skill that I've tried to develop now that my sons delight in the same sort of play. I have become more or less successful at it, although it works best if I have warning. My children are more likely to be able to tickle me if I have no time to marshal my defences. This experience lends weight to yet another theory about ticklish laughter: that it is simply a reflex. In Harris's article, she draws a comparison between the startle response and laughing in response to tickling. The startle response cannot be faked; it relies on unpredictability, and so, too, does ticklish laughter.

Yet there are those who actively search out and pursue tickling—both having it done to them and doing it to others. There is an array of sites on the internet dedicated to the erotic possibilities of tickling. A half-hearted perusal of them uncovers surprisingly tame images of dressed and semi-dressed women being tickled, usually by other women. Several feature participants rolled up in carpets, leaving their feet exposed to be stroked with feathers; other images ramp up the sexual frisson by binding the ticklee's hands and feet to bedposts. Other sites offer video images, promising such delights as 'lengthy underarm tickling' and 'shoe removal included'.

Each to their own. For me, the subtle shiver of knismesis far outweighs the dubious pleasures of shriek-inducing gargalesis.

THE IDEA that tickling and blushing, rather than technical dexterity or the complexity of language, might provide defining aspects of our humanity is immensely attractive to me. We tend to leave crimson cheeks and explosive fits of laughter behind as we grow older, perhaps because we work hard to grow a metaphorical skin—a kind of carapace—so that the blushes and giggles that reveal our emotions and vulnerabilities are rigorously stifled. Both responses—unambiguously physical and yet inextricably linked to our emotions—expose us, operating as they do on a level largely out of our conscious mind's control. They connect, paradoxically perhaps, to something that is close to the animal within us, and yet they are peculiar to humanity.

Unclean, Unclean: skin and disease

> 'The skin, after all, is extremely *personal*, is it not? The temptation is to believe that the ills and the poisons of the mind or the personality have somehow or other erupted straight out on to the skin. "Unclean! Unclean!" you shout, ringing the bell, warning us to keep off, to keep clear.'
> — Dr Gibbon in *The Singing Detective*

THERE ARE A LEGION of diseases and disorders that can affect your skin. Viruses, bacteria, parasites, allergies, drug reactions, congenital conditions, and cancer can all wreak havoc on your hide. Skin disease can turn your benign, elastic integument into a prison from which it is impossible to escape. Just ask anyone who suffers from a chronic condition that causes his skin to itch and burn. For anyone whose skin is in a constant state of inflammation, the desire to peel it off, even for a moment, must be intense.

According to Dr Josie Yeatman, a dermatologist, out of all of the medical specialties, hers has the thickest textbook. Qualifying to practise as a dermatologist, she tells me, takes a similar amount of time as training to become a neurosurgeon—a pertinent fact that she has had to point out more than once to new acquaintances who assume that she's a trumped-up beauty therapist.

Dermatology, she says, is able to be very specific about identifying and qualifying skin disease according to location and other factors, and this, together with the fact that there are so many structures within the skin that can be affected by disease and disorder, means that dermatology textbooks are very thick indeed. There are pictures, too—and they are not for the squeamish.

Your skin can erupt in warts, cysts, lesions, and boils. You might be plagued with verrucas, carbuncles, bunions, corns, or blisters. Pustules, rashes, pimples, and blackheads may ruin your complexion. Infestations of fungus or insects can result in scabies, tinea, or ringworm. Ulcers, tumours, melanomas, and carcinomas might threaten your life. Chronic disorders such as eczema, urticaria, dermatitis, and psoriasis can cause your skin to itch and weep. Highly contagious viruses like herpes, chicken pox, and shingles may signal their presence with unsightly blisters that seep and scab. The extreme sensitivity of skin, with its thousands of nerve endings, means that when it is inflamed it can be agonisingly painful and maddeningly itchy.

Skin can be affected by disorders, such as epidermolysis bullosa, where the slightest touch causes it to blister and peel. Children born with this disease are referred to as 'cotton wool babies', so carefully must they be handled.

Circus contortionists might owe their livelihood to Ehlers-Danlos Syndrome, where the skin's deficiency in collagen allows it to stretch to extraordinary lengths and then spring back to its usual shape.

Psoriatic arthropathy, a chronic hereditary disease, transforms the skin into a cracked, bleeding, flaking prison that is unable to regulate the body's temperature. Joints ossify, turning fingers into claws, and soaring temperatures induce hallucinations.

Dr Yeatman has dragged out the famously thick textbook and is flicking through it. The photos that she shows me are lurid and ugly: blistering and sloughing, ulcers, and gaping holes where skin should be. Later, when I listen to the tape of our conversation, it is filled with my sharp intakes of breath and horrified giggles cut short. Confronting images of just what can go wrong with skin jump out at me. A flick of the page and a newborn baby's face appears, deformed by a huge strawberry nevus.

'There are obviously big problems there,' Dr Yeatman remarks, 'The ear developing, the eye developing, feeding. It's a medical emergency, something like that.'

'This is nothing,' she continues with a turn of the page. 'We see this stuff every day.'

Where are these people, I wonder? I don't see this every day. It is rare that I encounter someone in the street with a truly terrible, disfiguring skin disease. When I pose the question, Dr Yeatman tells me that people who suffer these kinds of disorders are mostly reclusive, staying at home and surfing the internet or watching TV. Rather than suffering the discomfort of the disgust on the faces of others in the street, they elect to opt out of normal life. Others, though, brave the pain and the pity and forge on, determined to live a life unimpeded by the precarious condition of their skin.

For Dr Yeatman, the skin disease that she fears most is not Hailey-Hailey disease, where skin cells fall apart like 'a dilapidated brick wall' (a poetic description that I came across on a New Zealand dermatology website) causing terrible blisters; nor Sharpei Dog Syndrome, where the skin hangs in great folds so that sufferers tend to look like the eponymous hound; nor Behçet's disease, which results in painful ulcers in the mouth and genitals, as well as eye problems and skin lesions.

No, the skin condition that Dr Yeatman fears above all others is melanoma. The other, more ghastly, diseases are mainly genetic, and she knows that she is unlikely to get those; but melanoma can sneak up on person. The trouble with melanoma, a skin cancer caused by the uncontrolled growth of pigment cells, is that 20 years down the track, after apparently successful treatment, you can find yourself having an epileptic fit because a melanoma has popped up in your brain.

People are terrified of melanomas, and they are right to be. Public awareness campaigns have been so successful that your hairdresser is likely to pick out a suspicious looking mole and suggest you get it checked out. The risk factors for melanoma are well known, and include sunburn, especially when one is young; fair skin that is prone to burning; a family history of the disease; and having a large number of moles or having moles that are abnormal—those with an indistinct edge or of a large size.

Australia has a very high awareness of melanoma, to the extent that dermatologists now see people with mole phobia, and hundreds and thousands of benign moles are cut out of people's skins every year just to be on the safe side.

One in 150 people in Australia will develop melanoma; and the further north in the country that you live, the more common they are. Prevention is the best cure; but, if detected early enough, the cure rate is favourable. Advanced cases of the disease can see tumours spreading through the bloodstream to other organs including the liver, lungs, and brain.

In the pathology museum of New Zealand's Christchurch

Hospital, I saw a secondary melanoma that had been excised from the body it had threatened. Like the strata of an exposed rock, the different layers of tissue, including the uppermost layer of skin, were all clearly visible. The tumour was dark and toad-like, nestled between the subcutaneous layer of fat and the muscle beneath. Even in that state, suspended in a preservative solution within a large glass jar, the melanoma managed to emanate a pall of decay, and the threat of death.

In the same museum, cysts and other lumps and bumps of medical interest hang in solutions surrounded by the skin and flesh in which they'd embedded themselves. The written descriptions are usually brief, apt, and to the point: 'A well-defined cyst containing amorphous, yellow-grey material.' Sometimes, though, a more poetic turn of phrase is employed: one tumour with a variegated surface is described as having a 'watered-silk' appearance. Other words and phrases—if read by someone ignorant of the fact that they were describing a collection of cells with the potential to rapidly multiply and eventually kill their host—could almost be described as lyrical: ulcerating nodules, pigmented naevus, fungating melanoma.

Some of the growths are truly horrible and almost unbelievably large. One, the size of a small muffin, was cut from the thigh of a 55-year-old woman after it had developed within a year from a pigmented lesion that she had had since birth. Sometimes, the descriptions deviate from brief, clinical accounts to relate more information about the circumstances of their development. One explains the development of a mole, cut from a young man's calf, which had been struck by a ball during a game of cricket. In the following three years, it changed colour and steadily increased in size. Other tumours developed and, even with chemotherapy, he was dead a few months later.

Despite the very real danger of melanoma for fair-skinned Australians, the mother of all skin disease—the one whose name merely uttered can send a shudder of dread reverberating through our cells—has to be leprosy. Imbued with the horror of gross deformities and missing extremities, leprosy has a place in the collective consciousness as one of the most horrifying and hideous diseases of all time. Biblical epics, such as *Ben Hur*, have the obligatory scene that depicts rag-shrouded wretches extending deformed hands in a plea for alms (pun intended) while keeping their faces hidden, so terrible are they to look upon.

I have vivid recollections of gospel readings that stressed the obligation to ostracise lepers, and the notion that it was not enough to be healed of leprosy: one must also be 'cleansed' of it. Yet, despite this impression, which I'm sure I share with many people, if you type the word 'leper' into the search engine of BibleGateway.com (a website for reading and researching scripture online) expecting a multitude of verses to appear that detail the privations and rituals that such unfortunates have to undergo, you will be disappointed.

BibleGateway.com uses the *New International Version of the Holy Bible* as their default version of the Good Book, and you won't find lepers there. Rather, you'll find people who 'have an infectious skin disease'. Following in parentheses will be an explanation that the Hebrew word for leprosy was used for various skin diseases, not necessarily leprosy. So, rather than 'Command the children of Israel, that they may put out of the camp every leper' (from the *King James Version*), you will get: 'Command the Israelites to send away from the camp anyone who has an infectious skin disease'.

Now, call me old-fashioned, but to my mind 'anyone who has

an infectious skin disease' lacks the pithy punch of 'leper'. Lepers, and leprosy, for that matter, would seem to be out of fashion.

Rather than this being a clear case of too-stringent political correctness, a little bit of research reveals that leprosy has been somewhat maligned. (That is, if it's possible to malign a disease that can result in your nose rotting away.) While leprosy is infectious, it is not overly contagious; and, according to the *Encyclopedia of Skin and Skin Disorders*, 'patients should no longer be referred to as "lepers"'. Moreover, it is likely that leprosy didn't even exist in Moses' time, and so those hideous unfortunates from *Ben Hur* are figments of our collective imagination. Perhaps they just had a particularly nasty episode of psoriasis. The *New International Version of the Holy Bible* has obviously taken good advice.

Leprosy, or Hansen's disease, certainly does exist today, and untreated can be a thoroughly nasty affair resulting in blindness, nerve damage, and tissue destruction due to the body's reaction to the bacteria that causes the disease, Mycobacterium leprae. Nerve damage, especially in the extremities, means that the automatic withdrawal reflex is lost, as these areas can no longer sense pain. The result is more and greater injuries from hot or sharp objects, and scarring and loss of extremities such as the fingers and toes may result.

In the past, other society-sanctioned consequences were inflicted, including forcible sterilisation (due to an erroneous belief that leprosy was hereditary), confinement to a leper colony, or being made to rhythmically beat clapper sticks to alert others to your approach.

In the Bible's Book of Leviticus, lepers are instructed to tear their clothes, cover their faces, leave their heads bare, and to cry 'Unclean, unclean,' as a warning to others to stay away.

Leprosy has a five-year incubation period, and symptoms may take up to 20 years to appear. It is treated with multi-drug therapy, and early diagnosis and treatment can cure the disease and prevent disability. According to the British Leprosy Relief Association, over 610,000 new cases of leprosy were diagnosed worldwide in 2002, the vast majority of these being in developing nations in South East Asia, Latin America, and Africa.

Rejection and ostracism are still the norm for those with leprosy, despite the fact that early detection and treatment can result in a cure.

Whether it be leprosy or any other kind of infectious skin disease that causes your epidermis to weep, blister, crack, ooze, flake, fester, or rot away, diseased skin is ugly and an impediment to social intercourse. Skin, after all, as the doctor in Dennis Potter's highly successful television series *The Singing Detective* remarks, is 'extremely personal'. (Potter himself suffered from psoriatic arthropathy, and bestowed it on his protagonist Phillip Marlow.)

Despite the well-worn adage that beauty is only skin deep, appearances do matter. It is not too difficult to see an evolutionary angle to the value that most of us place on others' physical aspects. We would do well, in terms of our own immediate survival, to avoid those that exhibit signs of disease. Diseases of the skin are particularly repellent; weeping, red skin that sloughs and flakes or oozing white pustules invite few physical overtures, let alone embraces. The ideal skin is free of blemish, glowing, translucent, clear. Bad skin, whether it be because of acne, eczema, or scabies, is a source, for the sufferer, not only of discomfort but also embarrassment. The recoil is almost instinctive when you are confronted with the inflamed skin of another.

Our skin is out there, on display. A skin disease or disorder is difficult to hide and invokes a degree of disgust and discomfort that few other signs of disease do. Coughs and sneezes may be symptoms of disease, too; yet, while we may object to a spray of moisture from someone's mouth or nose on the train, cracked, bleeding, inflamed skin is likely to raise more alarm in a fellow traveller.

It's not entirely surprising that diseased or distressed skin can provoke such reactions. A friend who suffers from chronic psoriasis puts it like this: the skin is 'the barrier between inside and outside … it's breaking down, you're leaking out. It's disgusting.' The breaching of the skin, that which contains our body with all its messy fluids and pulpy viscera, is an abomination.

The idea of the 'personal' nature of our skin, which extends to an identification of the whole person with their skin, results in another phenomenon—a belief that the sufferer of a skin disease is somehow responsible for their suffering. After all, outbreaks of chronic skin conditions like eczema and psoriasis may be stimulated by stress. It is not much of a leap, then, to believe that if you can control your mind then you can control your skin disease. This is especially so, given, as we know, that from the embryonic level, our skin and our minds are linked, developing as they do from the same ectodermal origins.

Another friend, who has eczema, readily embraces the personal nature of skin, confessing that his skin is healthiest when he lives alone. 'When I've been in relationships,' he tells me, 'my skin's generally been pretty rancid.'

Talk of the skin as the boundary between our world and our selves leads him to speculate further. In romantic relationships, he says, your 'personal boundaries have been dissolved to include somebody else. My particular ecosystem seems to find that particularly challenging.' Yet suggestions that his disease, which can have him 'weeping, moaning, and railing at God', may be psychosomatic make him livid. The all-too-physical contaminants of dog and cat hair, or particular foods, are also likely to cause an outbreak, as can unhealthy living in general: too little sleep, too much alcohol, or not enough fresh food.

With a little too much candour for my comfort, he reflects on a particular relationship, which ended some time ago. (I remember the woman in question—a tall, elegant sylph.) Living in a domestic situation provided many circumstances out of which tension could arise, he says, and throwing skin irritation into the cocktail often exacerbated those tensions.

'Bella has sensational skin,' he goes on, dreamily. 'Her skin is so smooth. I would be fascinated just touching her skin.' Unfortunately for him, he tells me, frequent sex had the effect of making his eczema flare up.

The observations made by Andrew J. Strathern in his book *Body Thoughts* are consistent with my friend's experience. In some Melanesian cultures, Strathern writes, dry, flaky skin in men is attributed to a shortage of semen. This stems from the belief that semen is stored beneath the surface of the skin, and that sex saps a man's strength.

According to Strathern, how healthy, or otherwise, a man's skin might be is directly related to the state of his sexual relationships. Promiscuity or wantonness will result in 'bad' skin—a reflection not only of his physical health but also of his morals.

I'm not sure everyone who suffers from a chronic skin condition would appreciate such parallels being drawn between their skin and their moral state, but it is interesting to see that regard for the personal nature of skin is cross-cultural.

Chronic skin disease is often frustratingly resistant to treatment. Wilful and unpredictable, conditions like eczema and psoriasis may fail to respond to medical treatment, and then for no identifiable reason disappear—only to reappear without warning.

My friend who says his eczema is worse when he's in a relationship was advised by a worker in a Salvation Army op shop to bathe in his own urine, while an acquaintance told him sagely, 'It's definitely past-life stuff, definitely.'

My friend with psoriasis has tried many and varied treatments to get her condition under control, including cortisone creams of varying strengths, different herbal remedies, and vitamins. She has seen doctors, naturopaths, and a Chinese herbalist. Friends, relatives, and strangers have suggested calamine lotion, zinc cream, pawpaw ointment, and lotions derived from coal tar. ('Everybody's a doctor,' she complains.) A work colleague recommended a salve with the hyperbolic moniker 'Essence of Life'. While recommending the lotion as a treatment for inflamed skin, the label also extols its virtues as a plant fertiliser and car wash. The really alarming thing was that it seemed to work—in reducing the inflammation of her skin, that is; she's yet to try it on her car.

Finally, her dermatologist prescribed UV treatment three times a week. While she stands naked in a vertical chamber, a narrow band of UV light is shone directly onto her skin. The treatment, her dermatologist reassuringly tells her, 'hasn't yet been linked to skin cancer'. The UV treatment has resulted in some improvement, as well as a tan, but she still has psoriasis.

THERE IS A MARKED concurrence between skin disorders and psychiatric disorders. Given the prevalence of both, the tricky thing for health practitioners is to work out whether they are connected and, if so, what came first? Dealing with intense discomfort, lack of sleep, and the cosmetic impact of inflamed skin, sufferers of disorders such as psoriasis and eczema commonly also suffer from depression and anxiety. Dr Yeatman tells me that quality-of-life studies on psoriasis sufferers consistently find that the disease has a similar negative impact on a person's life as cardiac failure or cancer.

Psoriasis doesn't have a great public image either. It's not a very fashionable disorder. Celebrities will admit to drug and alcohol abuse before they'll put their hand up to poxy skin. Dr Yeatman told me that when she was researching questions for a quiz night for fellow dermatologists ('Which famous person has a particular skin disease?'), she searched the internet in vain for one Hollywood celebrity who would admit to having psoriasis. When you consider that 1 to 2 per cent of the population have psoriasis, she explained to me, there must be dozens of celebrities who have it. Yet the American Psoriasis Foundation can't get one of them to admit to it.

Conditions like psoriasis and eczema fall under what are known as psychocutaneous or psychodermatologial disorders because, while they are primarily regarded as having a physiological basis, they can be exacerbated by psychological factors such as emotional stress.

I remember the plump, redheaded boy who came to our small country school when I was a child. Even at our tiny school, where the prospect of any new kid who might provide a body to flesh out a softball or cricket team was keenly anticipated, he was immediately sentenced to social ostracism. His elderly parents, his carrot-red hair that flopped in front of his face, his large and numerous freckles, his inability to pronounce the 'r' sound, and his gentle nature all contributed to mark him as a victim. But nothing signalled his vulnerability to our hard-hearted little group as much as his habit of licking his lips. As a result of the constant nervous swipes of his tongue across them, his lips and the skin around them were constantly dry and flaking. It's a condition known as 'lick-lip dermatitis', and is a common disorder in children—a manifestation of a tic, or a habit, or, as I imagine in my classmate's case, the result of a nervous reaction to the cruelty of his peers.

Psychocutaneous disorders can be further categorised as being pyschophysiological, which is one of three broad categories of skin conditions that have a psychological dimension, and include conditions like acne and herpes. The other two are primary psychiatric disorders and secondary psychiatric disorders.

Skin conditions that are themselves a symptom of a psychiatric condition fall under primary psychiatric disorders. In a medical journal, a photograph accompanying an article on psychocutaneous diseases shows a matchbox filled with fluff and dandruff. It looks like the project of an obsessive toddler. A patient had presented it to their doctor as evidence of an infestation of parasites. Such behaviour is common in patients with delusions that their skin is beset with biting, blood-sucking beasties to the extent that it is referred to as the 'matchbox sign'.

Frequently, a patient suffering from such a delusion will have elaborate and detailed theories about the habits of the parasite, including its reproductive cycle, its movements, and its methods of exiting the skin. The caption cautions doctors not to dismiss the evidence presented to them via the matchbox sign. Before sending the patient off for psychiatric counselling, it is essential to ensure that there is no physical cause for the condition that he or she is complaining of.

Other photos in the same article show 'picker's nodules or excoriations', which are 'common dermatological presentations of anxiety and obsessive compulsive disease'.

The depression and anxiety that Dr Yeatman told me affects numerous patients who suffer from conditions like eczema and psoriasis, or disfiguring conditions like severe acne or vitiligo (loss of pigmentation), comes under the secondary psychiatric disorder. Studies have found that patients with severe skin conditions that result in cosmetic disfigurement have a much greater tendency to be suicidal than general medical patients do.

To add further complications to diagnosing a patient who presents with a skin disorder, some diseases, like liver failure or diabetes, may have both psychiatric and dermatological symptoms. Given the stigma that still attaches itself to mental illness, it is not surprising that general practitioners have to tread lightly and with tact if they feel that a patient who presents them with a matchbox filled with miniscule debris, as evidence of an infestation, might benefit from the treatments and therapies offered by a psychologist or a psychiatrist.

Perhaps the most common indication that something is amiss with our skin is itching. It is a curious phenomenon, that irritating, not-quite-tingle on the skin that demands to be scratched. Is there anything that feels as good, as satisfying, as right, as that moment when the slight pain of fingernails rasping across the skin perfectly intersects with that peevishness of nerve endings that we identify as an itch? For that fleeting instant, the sensation of the scratch and the itch combine to form something almost heavenly. There is also circularity to the itch–scratch response: scratching can cause an irritation that becomes itchy, and which is then scratched. One feeds the other in what can sometimes become a maddening and debilitating cycle.

A once-widely-held view was that the pain receptors located in the skin also transmitted the itching sensation to the brain. This was based on the observation that the mild pain caused by scratching an itch appeared to interrupt the itching sensation, and therefore relieved it. While it now appears that this is not the case—the itching and the pain receptors have been shown to be different shapes—recent research does indicate that pain and itching messages are sent to the same site in the brain, which happens to be where the emotions are located. Is it any wonder then that stress can have a significant impact on itching?

Philippa is a psychologist who spends part of her working life in a hospital pain clinic. Most of the patients referred to her have chronic pain that does not respond to the usual medications, or the side effects of the prescribed drugs that they take are onerous, and they are seeking an alternative to dealing with the pain. Occasionally, though, a patient is referred to her who suffers from chronic itching, either as result of a skin condition exacerbated by stress, or as a manifestation similar to phantom-limb pain.

Itching is very suggestive, Philippa remarks, and even as she says the word we both find our fingers furtively straying to scratch ... the place where the collar of a shirt meets the skin of the neck, beneath an ear, the underside of an arm. It's a phenomenon that I've often witnessed in the school playground: the ripple of scalp scratching that will radiate out through the waiting group of parents as they read the pink slip their child hands them to notify them that a classmate has nits.

In Philippa's experience, itching can often been seen as a reflection of one's emotional and internal processes. In assisting her patients to alleviate their itching, she uses similar interventions as those used for helping people to cope with chronic pain. The sad fact may be that the itching will always be with them, and her role is to help them to cope with a lifelong condition. Often this involves reducing the patient's arousal level through relaxation and meditation to help reduce the amount of adrenaline in their system. She also teaches them to recognise and prepare for those situations that they know may trigger itching.

SKIN IS PERSONAL, very personal. Diseased, distressed, or damaged skin, whether brought about by bacteria, genetics, or bad luck, won't win you any friends, and doesn't do a thing for your self-esteem.

My friend with atopic eczema recalls getting up from his desk after two hours of studying. His perpetual scratching had left a fine, white powder of skin cells surrounding his desk. 'Frankly,' he says, 'sometimes it feels like you are unclean.'

She's Got Perfect Skin

> Louise is the girl with the perfect skin
> She says turn on the light, otherwise it can't be seen
> She's got cheekbones like geometry and eyes like sin
> — 'Perfect Skin', Lloyd Cole

HER NOSE IS DELICATE and sweetly upturned. Traces of gold leaf adorn her skin, and her hair is henna red, evidence of her high rank. Even with her mouth closed, I can see that her teeth and jawbone are straight and well-formed—not because of the smooth fall of her cheek or the way her top lip meets her lower one, but because in places her skin is simply not there. Neither, for that matter, is the rest of her body.

She has been dead for 3000 years and, where once she would have enjoyed the privileges accorded to the Egyptian upper classes, now her remains are housed at the Melbourne Museum. Her head alone is all that is left of the person that was, and it lies unceremoniously in a wooden box, surrounded by other oddities and artefacts: a stuffed American beaver, a slab of flexible sandstone, a noughts-and-crosses playing machine. Granted the dubious honour of being one of the curators' favourite exhibits, and having being voted the same by patrons of the museum, she has been hauled out of storage for this special display.

Despite the skill employed to preserve her physical remains (a top technician would have laboured long and hard to remodel her nose after her brain was removed through it), her skin after all this time is blackened and shrunken onto her skull. Yet most of it is still there: after 30 centuries, her skin—that vulnerable, pliable body covering—still clings to the bones of her face. Helped along in no small part by skilfully applied chemicals and strictly controlled atmospheric conditions, it's true; but still, it is amazing to me.

NEVER, NOT EVEN in the dewiest flush of my youth, has my face possessed anything approaching the frozen, implacable beauty of the girl gazing out of the glossy pages of the magazine that I'm holding. She is lovely—ravishingly so. Her taut, tight skin, enclosing the bones of her skull in a silken sheath, makes the skin of my own face feel pouchy and lined, practically falling off my head in comparison.

'Resculpt your face to perfection' is the invitation offered by the cosmetics company using her features to tout its product. The claims are truly fabulous: a mere cream with the ability to 'shape and lift'?

The model in the photograph has to be a woman not yet out of her teens, and even her unblemished visage has, in all likelihood, been airbrushed, so alien is its perfection, so smooth the membrane of pores, glands, and tissue. The image is almost life-size, and the longer I look at it the more it gradually assumes a quality close to bizarre. The fall of her face, from her prominent cheekbone to her elfin chin, forms a line that is almost completely straight and at an improbable angle. An overzealous desktop designer has shaved a tad too much off from her natural contours.

Beauty has become an impossible goal, achievable only in a photograph and then requiring the careful manipulation of pixels. But it is the expression on the model's face, or rather the lack of it, which truly marks it as not-of-this-world. Framed as it is, its only function is to be gazed upon. It is a face impervious to emotion, untouched by the merest hint of a smile or the mildest puckering of the forehead in thought, let alone anything so animated as laughter, the sudden, sharp gasp of surprise, or a crumpling moan of dismay. Its state of being is one of vacuity. It is a void without spirit or spark, as vapid as the face of a store dummy.

This is beauty as defined for us today, and advertisements like the one before me implicitly promise that we are all able to attain it. It's waiting for us, buried deep within; we only need the courage and the cash to reveal it. The skin, they imply, is merely the raw material: let us value-add for you. One only has to commit to a cream, a moisturising wash, an injection of toxins, an abrasive mix of chemicals and crushed rock, or the deft flourish of a surgeon's knife.

In case it's slipped your notice, fresh, unlined skin is prized, representing as it does beauty, wealth, and youth. An entire industry is based on the desire to keep our skins as pristine as possible. Beauty may only be skin deep, but it is in the skin that it begins, that it has its bedrock. Finely modelled cheekbones, lustrous hair, and a pixie nose will take a girl a certain way, but without the delicate bloom of smooth, clear, radiant skin, what woman can hope to be truly exquisite?

We buy sunscreen to prevent skin damage, apply moisturisers to promote elasticity, and lather on anti-ageing creams to repair the ravages of time. Some subject their skin to chemical peels in order to burn off the outer layers to reveal younger, fresher-looking cells beneath. Others inject toxins to paralyse facial muscles to inhibit frowning, or fillers like collagen to plump out creases and wrinkles. There is little discussion about where these substances come from. Collagen is usually derived from cattle tissue, but a disturbing investigation by the *Guardian* newspaper in 2005 discovered that a Chinese cosmetics company had been harvesting the substance from executed prisoners and aborted foetuses for sale in Europe.

Such gruesome realities haven't dented the cosmetics industry, however. Every celebrity worth her implants seems to be flogging some kind of skin-care product, from facial whiteners to creams containing antioxidants and vitamin C. Of course, it doesn't come cheap—not for those of us who just buy the stuff, nor for those creatures from whom it is harvested.

Beauty has a long history of being associated with innate virtue—think Cinderella or Snow White. In recent times, this has been extended to include the view that keeping oneself as physically attractive as possible is practically a duty. It's almost considered virtuous to look younger than you are. Perceived neglect of one's appearance is viewed with disapproval. Celebrity beauties are applauded not for their achievements in their chosen field but for the way that they continue to look 35 while their calendar age marches relentlessly on to 50 and beyond. Faced with these endless advertisements of smooth-faced beauties, some who are my age and older, it's easy to feel that soon I will be the only woman on the planet who looks middle-aged.

Of course, we are not the first people in history to consider the appearance of ageing to be undesirable, or to try to develop remedies for wrinkles or strategies to prevent them forming. The long-dead Egyptian aristocrat now in residence at the Melbourne Museum may well have used, as a remedy for wrinkles, lotions containing milk, wax, olive oil, incense, hippopotamus fat, and gazelle dung before her death.

The Victorians had a range of lotions and treatments for wrinkles, some less benign than others. Ladies might repair to bed at night with masks of suet, raw beef, and wet cotton to ensure their faces remained unlined. Compounds of lead and mercury, or washes that included corrosive sublimate, prussic acid, or arsenic were also employed to keep a fresh complexion. One treatment recommended in the 17th century reads almost like a religious rite: powder of myrrh was placed on a shovel held over a fire and then sprinkled liberally with wine. Taking a mouthful of the same wine, one would lean over the smoking concoction, allowing the fumes to billow around the face.

IN THE COSMETICS DEPARTMENT of the retail store David Jones, the stark white walls have the same featureless, blemish-free aspect as the skin of the 'perfect' face. They also manage to convey overtones of a hospital or medical clinic—it's science we're dealing with here, not simply sweet-smelling unguents—according ageing skin the status of a disease.

Yet there is also the whiff of the old-time spruiker on the back of a wagon: 'Roll up, roll up. Get yourself a bottle of Dr Peabody's Ubiquitous Tonic: efficacious for flaky skin, epidermal eruptions, and inflammation of the bowel.' Brochures fanned across the glass cabinets where fabulously priced lotions are stored pose rhetorical questions: do you suffer from Chronic Silent Inflammation (note the capitals) caused by *everyday life*? Did you know that when your skin suffers an *assault* (we're talking sun, wind, and stress here, not a knife-wielding thug) it triggers a *state of emergency*?

Admittedly, today's 'cosmeceuticals' aren't all snake oil and essence of violets. Some contain substances that have been proven to alleviate the signs of ageing. Retinoic acid, for example, which is derived from Vitamin A, can reverse some sun damage if used long-term. Applied directly to the skin, it sloughs off the outer skin cells, reducing the appearance of fine lines, and it may increase the amount of collagen in the skin. However, it can also cause irritation and make the skin more susceptible to sunburn. Skin cells replace themselves at a slower rate as we age, and alpha-hydroxy acids, derived from fruit, if used regularly and over time, can also encourage the outer layers of skin to shed, and so expose the new fresher-looking skin cells beneath.

Back in the gleaming cosmetics department, I pick up another brochure that claims 'ageing is optional'. The formula that is the basis of the products it promotes has been 'discovered, researched, formulated, patented, and tested' by a doctor whose name is the brand. I gawp at a small bottle with a price tag of $900. The advertising material claims that the cream has been developed using 'Nobel prize-winning technology'. There's a part of me that wants to believe the claims that my skin can be rejuvenated, that ageing can be postponed, that the lines etched into my forehead can fade.

One bottle has been labelled 'tester', so I feel that I have permission to dab some on my face. As I screw the lid back on, a fragrant presence asks, 'Want someone to play with?' The speaker is male, young, and gorgeous. In this shrine to the possibility of feminine perfection, his masculinity helps to render him relatively non-threatening. Yet I have to suppress a thought that he could be the Devil. I have a vision of a pint-sized angel appearing on my right shoulder, urging me to walk away. But his invitation has been couched to defray my alarm and scepticism, and I let him dab more of the incredibly expensive stuff on my face. His patter is as gentle as the mini-massage that he's giving me, and he trills the phrase 'neuropeptide facial conformers' without tripping over it. These 'conformers' work on the muscles and transmitters in the skin to tighten it, he tells me. He explains that the product he's putting on my face is what one would choose before going down the path of Botox or surgery.

I like him—he is kind: 'I'd just recommend the eye cream for you; elasticity is not a problem at your age.' My skin tingles not unpleasantly.

He directs me to a mirror. Am I mad or is this product actually working? The skin around my eyes and mouth seems less lined. He chides me gently for putting some of the face cream on before he got to me. Otherwise, he says, he would have put cream only on one half of my face so that I could really see the difference. He asks me about my skin care 'regime', and I mumble half-truths about never using soap and moisturising twice a day.

Part of me wants to succumb to the ministrations of this skin-care angel/devil offering the promise of eternal youth, but I know, and he knows, that I'm not going to spend that sort of money on this stuff. I take my tingling face away.

A FRIEND'S SIX-YEAR-OLD DAUGHTER is given pencils and paper to keep her occupied while her mother and I catch up over a cup of tea. The child announces that she is going to draw our portraits, and sets to work. As she nears completion, she pauses to compare her efforts with the real-life versions in front of her.

Her smooth brow puckers with discontent as she looks at her drawing and then back at her mother and me. The likenesses are unsatisfactory, it seems; something is missing. Her face suddenly clears: 'I've forgotten your stripes!' With her pencil held like some sort of gouging tool, she ploughs a series of horizontal lines across our portraits' foreheads.

WHERE DOES THE SOFT, PILLOWY SKIN of a baby go? The plump, fresh cheeks, the firm, unlined brows, and the smooth, clean limbs of children contain a beauty and integrity that they wear with complete indifference. Their bodies are themselves, and they waste no time gazing at them objectively. Adolescence begins the process of separating the body from the self, and as we grow older, it seems, the body, and the skin it is encased in, lets us down. Nothing betrays our age with more unrelenting rigour and pitiless exposure than our skin. Our youth-obsessed culture insists that wrinkles suck beauty into themselves, and there it rots and withers away.

As we age, collagen and elastin fibres that previously provided flexibility and support start to thin. No longer able to spring back into shape, our skin begins to sag and bag. Frown lines and laughter lines slump into permanent furrows; capillaries weaken, making us more prone to bruising; and blemishes appear—brownish-yellow keratoses, age spots, and small cherry angiomas. Our skin cells take longer to regenerate, wound healing is slower, and the outer layer of the skin becomes dryer and rougher due to reduced sebum production. The melanin-producing cells decrease in number, our skin colour becomes patchy, and blotches develop.

Where once our skin cells lined themselves up in tidy rows like well-behaved schoolgirls to form a smooth complexion, they now begin to jostle and meander, allowing the skin to fall into folds and wrinkles. Creased bags of skin begin to develop under the eyes, while deep furrows appear either side of the nose and running down to the mouth, lines ripple across the forehead, and the skin of the cheeks droops so that jowls hang under the jawbone. Lines extend down from either side of the mouth, giving the impression of permanent glumness, and crows' feet radiate out from the corners of the eyes, at first only appearing when we smile, but then softening and deepening into permanency.

Coarse hairs begin to spring up—in women, on the chin and upper lip; in men, in the ears, nose, and from the eyebrows—as if to compensate for the thinning hair on our scalp. Subcutaneous fat levels decrease, particularly on our faces, shins, hands, and feet, but are likely to increase around the stomach and thighs. And what's wrong with all that? you might ask.

If we don't end up with the face that we deserve, then perhaps, finally, we end up with the face that we couldn't avoid. But can't beauty exist at all stages of our lives?

Youth has a simple and uncomplicated beauty in and of itself: experience has yet to imprint itself onto the unlined face of a child; their beauty is all about potential. Some faces, though, grow into their beauty, shaped by life and experience, reflecting qualities and aspects that are unique to the people that they front. Laughter lines may simply be crows' feet when you're 45 and looking in the mirror, but these furrows and wrinkles can convey warmth, a life fully lived, or strength of character.

Going through a book of photographic portraits—some of the rich and famous, some of the plain and anonymous—I look for wrinkles and for beauty. More often, it is the men caught in these photographs who seem to wear the weathering of their skin lightly. Yes, writing it makes me feel like a traitor to my own sex.

For Spencer Tracy, for instance, a face full of wrinkles can't mask his star quality or his sex appeal. Mel Gibson, no longer a youth, but not yet the self-confessed alcoholic with a hoary beard and a propensity for anti-Semitism, is uncomplicatedly handsome with the lines of maturity beginning to etch his face. A photograph of an impossibly young Mick Jagger makes me immediately think of the worn-visaged grandfather of rock that he is today. Mick hasn't had plastic surgery, although it's obvious that his vanity is still intact.

There is shot of Vivien Leigh with her husband Laurence Olivier, taken in 1951. It is a close-up—she glances up at him, her brow ever-so-slightly wrinkled and with fine lines around her eyes. She is a mature woman here, not the pristine beauty that she was in *Gone with the Wind.*

I turn the page, and suddenly there before me is a photograph of a 16-year-old Elizabeth Taylor—flawless, luminous, unlined, young. Suddenly, Spencer Tracy and Mel Gibson can go to hell—here is beauty, pure, and simple.

THE SKIN CLINIC is in a traditionally working-class suburb of Melbourne that is in the process of a steady gentrification. While house prices are steadily rising here, it's hardly the haunt of well-heeled matrons with too much money and too little time. Several shops in the same block are vacant but, a few doors down, a pawn shop on a corner site appears to be keeping its head above the fiscal waterline.

The clinic—and it calls itself a clinic rather than a parlour or a salon (both terms that have assumed a quaint 1950s aura)—exists in a little hub of self improvement among the empty buildings, pizza shops, 24-hour convenience stores, and laundromats. It shares premises with a 'college' promoting meditation and healing, next door is a fitness centre, and across the road is a hairdresser that offers therapy for hair.

I imagine women walking into these establishments with the cash that they've just obtained from pawning their grandmother's engagement ring to pay for a Botox injection, a hair rejuvenation treatment, or a crystal-healing session—swapping their families' heirlooms that so they can invest in themselves.

The reception area is painted in soothing lilac tones; scented candles, glass bowls, and coloured pebbles are arranged in tasteful combinations. Crystals in various forms are displayed for sale. The staff members wear white tunics, giving them a brisk and efficient look that telegraphs the carefully crafted message of therapeutic, rather than cosmetic, intervention.

Most of the treatments offered include the word 'medical' in the title: it's a strategy that's a little too obvious and not quite successful, especially on brochures that promote Tibetan healing alongside Botox.

It's as if the clinic itself has a multiple-personality disorder: part science, part affluent hippiedom. There is little mention of youth or beauty; the context is self-improvement, even self-realisation. Obviously, the word vanity does not get a look-in. The skin-care industry has elevated such things as freckles, large pores, and laughter lines to conditions that require treatment: your skin has needs. Skin rejuvenation is not so much an option as an obligation—to yourself and your partner, not to mention to the rest of the world, who otherwise will have the unpleasant experience of looking at your tired, furrowed face.

Assuming that the staff are, to some extent, advertisements for the services offered, I check them out. One has glabellar frown lines like mine, and another has a comfortably big bum. It's both disappointing and consoling: at least they're not all uber-models; and yet the implicit promises contained in the brochures seem to demand something close to perfection in the purveyors of these services.

Soon, I am called to the consulting room, which is small with walls painted a medical white. One wall is a mirror, which gives the impression that the room is bigger than it is, and allows for close examination of one's imperfections. On the desk, alongside a conventional computer screen, is an electronic box similar to a large PC monitor but with the screen and wiring taken out. The front of the box is open, and inside it are lights and little mirrors that beam down towards a chin rest.

On instruction from the therapist, I place my head in the required position, close my eyes, and wait for the blue flash that tells me that my photo has been taken. The rests for my head are moved, and the other side of my face is then recorded. I can see myself reflected in one of the little mirrors inside the box and, this close up, it isn't pretty.

'You're …' the therapist scans the sheet where I've filled in my details '… forty-four. So you're getting to an age where you want to start looking after your skin, so that by the time you're 50 or 60 …' She pauses, looking at me. I resist the impulse to fill in the gap by saying something self-deprecating, and wait for her to dig her own verbal grave. Eventually, she gives a little half-smile and leaves the sentence unfinished.

I look at her skin. Not perfect, but good. I can see some little bumps under the surface, but on the whole it's smooth and the pores across her nose are barely visible, unlike mine. She is younger than me by 10 years at least, and her demeanour is familiar and just a little patronising. She calls me 'my dear', which I dislike only mildly.

The computer screen is turned towards me so that I can see the series of images of my face generated by the white box that swallowed my head. It's worse than I thought: this close up, my face looks like that of a bog person pulled from the peat after lying undisturbed for 2000 years.

The therapist clicks the mouse to zoom in on one of the images, and coloured, squiggly lines appear, circling the blemishes on my face so that it resembles a topographical map—all contour lines and shaded areas. Each image highlights a particular flaw: sun damage, visible capillaries, enlarged pores, oily areas, dry areas, built-up skin cells, blackheads, scars that I didn't know I had. She points them all out with an almost-motherly matter-of-factness ('Who'll tell you if I don't?' is the unspoken justification for this frankness that borders on the impertinent).

The evidence of my neglect, poor skin-care regime, and profligate ageing is all too apparent. I have been recklessly prodigal in the treatment of my skin; that much is clear. I feel that I need to apologise to someone for my rampant irresponsibility. Seeing these close-up images of my pitted, uneven complexion, I'm prepared to sign up for just about anything: I just want to look less hideous. Every time the therapist mentions wrinkles, she rubs her fingers over her own face, mirroring where my own most noticeable lines are on my forehead and between my eyebrows.

The therapist explains the treatments that she'd recommend for me: medical microdermabrasion that uses fine, crystal granules cut into the shape of diamonds (merely for the glamorous sound of it?), and medical intense pulsed light treatments. She stresses the word 'medical' whenever she uses it. The treatments would be staggered over 17 weeks, and even her promise of the brown marks coming to the surface and flaking off seems plausible. Goodbye to my congested epidermis. My skin will regenerate faster; the appearance of lines will be reduced. It sounds reasonable, achievable, desirable—necessary. I want to be medically improved, to place my face in the hands of experts who know and care about these things, to come out the other end of the process smooth faced and small-pored with younger-looking skin.

Then she mentions the cost. The figure makes me swallow hard. She must have noticed the non-surgical tightening of my throat muscles, because she hastens to assure me that it's very good value. I guess it's a matter of definition: for the same money, I could fly to Rome and back, and pay for a week's accommodation. I could take care of both my children's music lessons for the next two years. I could buy myself a red leather couch. I don't even want to think about what it might do for an African village.

The therapist follows up with a folder of before-and-after shots of other clients. I'm relieved: they're a little underwhelming. Yes, I can see a difference: brown marks faded, redness reduced, lines lessened; but, except for the pair of photos illustrating the successful treatment of visible capillaries, the results are not astounding.

I let out a breath: I can live without medical microdermabrasion and medical intense pulsed light treatment. I tell her that I'll think about it, and get out of there. Still, the skin on my face now feels as thick as buffalo hide.

I walk out the door of the clinic, and straight into a pharmacy a few doors down, and begin scouring the shelves. A fresh-faced assistant approaches, and I announce that I want an exfoliator. She shows me what they have, and recommends a particular one that she uses. I'd need about 50 years' worth to come close to the same price for the treatment outlined by the clinic. I look at the pharmacy assistant's skin. It's good—almost as good as the therapist's. I buy the little bottle of cream with abrasive particles in it.

On arriving home, I go to the bathroom and scrub my skin. My newly abraded face looks back at me from the mirror. I notice a slight difference—I do. My skin feels smoother, my pores are smaller, the lines on my forehead are slightly reduced—truly. Tracing my fingertips over my face, I practise tightening the skin of my brow to flatten out the already-permanent lines on my forehead. I really have got to stop frowning; it accentuates the furrow between my eyebrows and deepens my wrinkles. I think back to the head of Egyptian princess. Now, where can I get my hands on some hippopotamus fat and gazelle dung?

The Lure of the Tattoo

> I always look for a woman who has a tattoo. I see a woman with a tattoo, and I'm thinking, okay, here's a gal who's capable of making a decision she'll regret in the future.
>
> — Comedian Richard Jeni

THEY ARE EVERYWHERE. My hairdresser has at least one. So does the girl at the Myer makeup counter: the eye on her wrist squints at me as she brushes over-priced foundation on my face. The buff young baker at the fashionable café has tiger stripes the length of one arm, and they look so impressive that I let my mind wander as to where he may have more.

Two of my three sisters have them. Celebrities are lousy with them. The Australian army, in 2007, lifted their restrictions on enlisting people with them.

They wink at you from above the waistbands of nubile young things' jeans, brazenly drawing your attention to the skin below the belly button and beyond; or they peep seductively from the neckline of a low-cut top. They dally beneath the straps of young mothers' floral sundresses, or boldly declare the name of a loved one in a profusion of flowers and hearts.

On the tram, an older man has what look like self-made ones

fading into the skin of his forearms. Across the aisle, a pretty young woman has BABY DOLL etched in ornate lettering across the knuckles of both hands, prompting the shrew in me to think, 'You look sensational now, Baby Doll, but will the irony be intended when you're a tired forty-five?'

I will declare myself now: I am no tattoo aficionado. Most tattoos strike me as garish and, forgive me, ugly. Clichés abound, and elegance is largely absent. And tattoos leave such large room for regret; it doesn't really matter who dumped whom, you're still left with their name under your skin.

Only once have I seen a tattoo that I coveted for myself. It was on the upper arm of a young woman whom I passed one summer evening in St Kilda. A mermaid—not the garish fantasy creature favoured by sailors but a fey suggestion of salt and sea—existed in a happy confluence of deftly drawn lines. A whimsical creature, half-woman, half-fish, she hovered on the girl's arm; a blue, curved line for her belly, waves of parallel lines for her sea-swept hair; a series of small arcs for the shimmer of iridescent scales. She had a feminine mystique that was unsullied by the fuggy fantasies of fishermen. What clinched her appeal for me was the way in which the tattoo artist had incorporated a small red nevus on the woman's arm into the design to form the mermaid's navel. The tattoo had become enmeshed with her skin in an organic way.

It was in St Kilda, too, that I saw a tattooed number on the forearm of an elderly man. I was a naive girl from Queensland, and I still remember the quiet jolt of shock when I recognised its import.

For a short time, I tantalised myself with the idea that I might get a tattoo. Saying it out loud gave me a thrill at my own daring. Should I get the copyright symbol of a c in a circle inked onto the inside of my forearm? (I'm a writer—get it?) Or perhaps, like an Arabian woman from an earlier time, I could have three dots tattooed in a triangle on the palm of my right hand to ensure the continuing love of my husband. The thought of a needle puncturing the sensitive skin on the underside of my paw was enough to make me twitchy, but I liked the simplicity of the design with its nod to the arcane.

In search of a tattoo that I could bear to have, I scoured books celebrating the art form; but, in the end, I found that they just depressed me. Even the glossy volumes with high production values were unable to elevate ink-on-skin for me. The designs were garish, the colours bilious. Gothic images of blood and hearts abounded, as did tired, humourless portrayals of big-breasted women and empty hyperbolic statements rank with bravado, such as 'death before dishonour'. Several times, I saw the 'classic' hunt scene of dogs and scarlet-coated horseman weaving across the landscape of someone's back in pursuit of a fox disappearing into the cleft of the buttocks. Hilarious—once.

In one memorable photo (oh, how I wish it wasn't), a man lay naked on his stomach, his rear end towards the camera. His legs were spread wide, so that the view was of his buttocks and scrotum; a snake emerged out of his anus, and curled over his backside. He must weary of having to assume this position every time someone says, 'Go on, show us your tattoo.'

The commentary in these books did little to inspire me, either: a caption under a photo of a plump woman displaying the ornate tattoo on her back read, 'You don't have to have a beautiful body to have a beautiful tattoo!' The jolly exclamation mark made me want to puke.

It's true that some of the historical images had a dignity and an authenticity largely absent from the majority of the more recent tattoos displayed in these volumes. The Maori moko, which adorn the full faces of the men and are chiselled into the chins of the women, has gravitas within its swirls and spirals that is unmatched by bluebirds and lucky horseshoes. Cook's voyages in the Pacific, where the sailors were intrigued by tattoos on the skins of Polynesians, are often credited with inspiring the cult of tattooing among sea-faring types, and with introducing tattooing to Europe; however, the Picts and the Celts had been tattooing themselves before Roman times.

Now, in Western urban society, it is as unremarkable to have a tattoo as it would have been in Aotearoa before the coming of the Europeans. What once was the mark of the criminal and the outcast is now the badge of the young and the hip. Recently, 'arse antlers', referring to a tattoo just above the buttocks with a curving extension on each side, received its own entry in the *Macquarie Dictionary*.

No longer exclusively confined to the men of the working classes, tattoos are now found on doctors and lawyers of both sexes. As a symbol of rebellion and individuality, the tattoo has been well and truly devalued—Lachlan Murdoch, elder son of media mogul Rupert Murdoch and a member of Australian business royalty, has one, for god's sake—although there are still ways to use a tattoo to outrage and discomfort. Facial tattoos, for example, will most likely ensure that you have a seat to yourself on public transport, while the crude, amateur marks of homemade tattoos won't win you any awards for fashion.

Obscenities might amuse your friends for the first week or so, but you'd be advised to cover them up for a job interview. On the other hand, irony is acceptable, and so is homage. Quirkiness is definitely okay, Eastern and tribal imagery are practically de rigueur, and so are dainty images for women: delicate little fish, a ring of flowers around the ankle, a bluebird on the shoulder.

Facial tattoos remain particularly confronting, and most tattooists will refuse to do them. Tattoos on the face inevitably impose themselves in between personal interactions. Chopper Read, the self-confessed killer and now painter and children's book author, is not short of a tattoo himself, but his bodyguard goes one up on his boss. Tony 'the Face' Cronin works the tough-guy thing as far as it will go with a full facial tattoo. His dark sunglasses, close-cropped hair, black clothing, and flint-faced demeanour amp up the intimidation stakes, too; but it is his full-face tattoo that signals 'hard man' more than anything else.

Tattoos on the hands, like those on the face, also transgress the shifting boundary of what is objectionable. Like facial tattoos, those on the hands are difficult to conceal, and play a significant role in social interactions. Wearing gloves—except in cold weather, or as protective apparel, or perhaps to match your gown at the Oscars—is no longer commonplace. Ever since Robert Mitchum played the murderous preacher in *Night of the Hunter* with LOVE and HATE inked across his knuckles (prefacing BABY DOLL by half a century), the motif has been echoed in films and television from *The Blues Brothers* to an episode of *The Simpsons*. There is something about letters on the knuckles, perhaps because they are displayed most effectively when the hand is curled into a fist, that signals violence and anarchy.

SHE LIES NAKED on a froth of flimsy material, her chest propped up by a cushion that modestly shields her breasts from the gaze of the camera. A half-smile hovers on her lips; her hair is a blonde tumble. Her legs are bent at the knees, with her feet in the air to display the tattoos that twine around her calves and feet. Butterflies float on her lower back, and more tattoos meander up towards her shoulders. On her right upper arm, a naked nymph sits astride an elaborate, mythically proportioned fish. Her lower arm is a welter of flowers garlanding a dragon-like creature. On her upper left arm, a topless woman with a carelessly strung lei and a short grass skirt leans against the trunk of a coconut palm, while a snake writhes around a dagger that adorns her lower arm. There are other images inked onto her body—a grave marker, a serpent, and a dagger—but the photograph is in black and white, so the colours have to be guessed at. She is Cindy Ray, 'the classy lassie with the tattooed chassis'.

Bev Nicholas (formerly Bev Robinson) adopted the name Cindy Ray when she went on the road in the 1960s as 'the girl who put the oo in tattoo'. One of the first heavily tattooed women in Australia, her name comes up again and again in books about the history of tattooing; she is an icon in body-art circles. She still lives and works as a tattooist in Williamstown in Melbourne, where she was born.

When I arrive at Moving Pictures, the studio that Bev ran for 35 years and still works in part-time, I'm greeted by an incessant buzz, like that of an enraged hornet, drifting out the open door and down the footpath. Bev's client, Dennis, is having an old tattoo gone over with a new one—a bunch of flowers. It may not sound very butch but, with the aesthetic and lurid colours of an old-style tattoo, not to mention the words 'love' and 'hate' tattooed on his knuckles, Dennis can more than pull it off. He can't complain, not out loud, anyway—he's had an operation to remove his voice box. 'He's the silent type,' says Bev.

This is the first tattoo that Bev has done in over a month: illness in the family has kept her busy and preoccupied. She's well practised, though, and the hiatus hasn't made her hesitant. Mind you, she has been tattooing for over 40 years. She's confident and quick, drawing the outline of the design with steady hands, holding Dennis' wrist fast while she works on his forearm with the tattoo needle. Bev's wielding of the needle is by no means rough, but her technique is firm and steady rather than gentle. Dennis' arm is quivering slightly, and Bev notices.

'It's a bit painful up here, Dennis,' she says, as the needle works over the bony part of his arm just beneath the elbow; but she doesn't stop. 'You're turning 60, aren't you, Dennis?' she asks to distract him.

He nods.

'It'd be lovely to be turning 60 again,' she says, half to herself. Every now and then she stops the needle and leans back, tilting her head slightly to get the overall effect of her work. She wipes the skin regularly to remove the excess colour and the small amount of blood that oozes, and then sprays a disinfecting solution over the area. After wiping it down again, she smears it with Vaseline, allowing the needle to slip more easily across the surface of the skin. Then the buzz starts up again and the needle is running smoothly, tracing swirls and curlicues in gorgeous colour.

Bev is dressed in khaki pants, a black T-shirt with a tiger motif, and yellow rubber gloves. The skin between them and the sleeves of her T-shirt are covered with ink on both arms. She also has small tattoos on the lobes of both ears. Her long, straight, silvery hair is pulled back in a ponytail, and her fringe is a straight line across her brow. She wears glasses while she works, but her eyes are bright and her face has the open, quick-changing expression of someone who engages with life despite sorrows and setbacks.

Bev's not sure what it was that convinced photographer Harry Bartram that she would be up for his money-making scheme all those years ago or, for that matter, why she let herself be talked into it. He spun the then 19-year-old keypunch operator a tale of the fortune that she could make as a tattooed woman. Despite the fact that no one in her family had tattoos, and that getting tattooed 'just wasn't done', she got four on the first night. Her parents were disgusted although, some time later, her mother did allow Bev to practise her tattooing on her. 'My father hit the roof,' Bev recalls, and the little duck tattooed on her mother's arm was hidden forever.

After tossing in her life on the road, Bev set up the studio in Williamstown. It's old school: the walls are hung with tattoo designs, many the creations of Bev's ex-husband, who taught her the trade.

Tattooing may once have held the glamour of the transgressive, but any such allure has long since faded for Bev. If she'd stayed working as a keypunch operator, she says ruefully, she'd be living off a decent pension, and not sitting here in the shop waiting for customers that may or may not turn up. She sold the business some years ago, but Kenny McPharlane, the new owner, begged her to keep working on a part-time basis. 'Everybody's asking for you,' he told her.

After Dennis leaves, a younger bloke comes into the studio. He's already fairly heavily tattooed, and obviously knows Bev, at least by reputation, because he calls her by name. He explains that he'd like the pair of lips he had tattooed on his buttock covered over with a star similar to the one that's already on the back of his neck.

How big are the lips? Bev wants to know.

The young man shrugs, 'Not very big.'

Bev gestures for him to drop his daks so that she can have a look. We're standing in the front of the shop, the door open to the street, the big plate-glass window inviting the glances of passers-by. I'm not sure what the etiquette is here, but Bev obviously isn't about to draw a curtain, literally or figuratively, to pander to the young man's modesty and, after a slight hesitation, he pulls down his shorts to expose the bright red lips on the side of his right flank.

'What the hell made you get lips on your bum?' Bev asks rhetorically, bending down to get a closer look. 'They're not small, are they?' she says, looking up at me for confirmation.

I glance down at the tattoo under discussion. 'They're life-size, I guess.'

Bev gets out the tape measure. She hasn't got a star stencil that's big enough to cover them, and she suggests that he choose something tribal instead. He yanks up his shorts and begins to browse the designs on the wall.

Bev was inducted into the Lyle Tuttle Tattoo Art Museum's Tattoo Hall of Fame in St Louis, USA, in 2005. Not that she goes in for that sort of stuff much. She was gracious enough to attend, but all that attention is not really her cup of tea. 'I don't like people making a big fuss of me,' she says.

Kenny reckons that she's probably more famous internationally than she is in Australia. 'She's an icon. We get a lot of tattooists from all over the world coming in just to meet her. I think any self-respecting tattooist would know who Bev was.'

Bev's still self-conscious about her own tattoos, and will cover up if she's out somewhere. 'I've always got something in the boot of the car with long sleeves, no matter how hot it is.' She thinks it has a lot to do with her late father's attitudes to tattoos. 'He was always saying, "Put a long-sleeved jumper on. Don't let anyone see them."'

Back then, the brazenness of the young women today, who casually display their large and elaborate tattoos, was unthinkable. 'They're gamer than me,' Bev says. She admires many of the more contemporary styles of tattoo: 'If I did decide I was going to get tattoos done again, I wouldn't go for all the coloured stuff. I'd go for the wash work [a style of tattoo that uses only black ink]. I think that looks really lovely.'

Bev got her last tattoo in 2007, her first in 17 years. She pulls up the leg of her tracksuit pants to reveal a simple design of just three words above her knee: 'My Son Craig', and beneath it the dates of his birth and death. He passed away in July that year after a long battle with cancer. 'I fell in a heap,' she says recalling her loss; but she continued her regular Sunday tattooing shift at Moving Pictures. '"At least it's getting me out of the house one day a week," I thought.'

A few months later, Kenny rang her with a warning, 'Now, don't yell at me when you walk in the shop. There's a surprise down there.' The surprise was a huge mural on the back wall of the studio, based on a photo taken of Bev when she was Cindy Ray. It shows a young, pretty blonde woman, a shawl draped modestly around her shoulders, with a winged woman tattooed on her upper chest.

Kenny commissioned the mural as a tribute to Bev—to pay his respects and to ensure her name and reputation carry on. 'It's probably been the highlight of my tattooing career to have met Bev and to have actually worked with her. To buy her studio was just the icing on the cake.'

There's just one more thing Kenny would like to do: 'I have to get a tattoo done by Bev. That's a must.'

TONI MEETS ME at the door. Her baldness is so complete that she almost gives the impression of being a cartoon character. Her pate is smooth, her eyebrows and eyelashes absent. The chemotherapy has done its work for the moment, killing off the fast-growing cancer cells, but her hair has been the other casualty. Her head looks vulnerable in its nakedness, but her strength is evident, too. 'I've embraced my baldness,' she tells me, recounting that while in Byron Bay recovering from a bout of chemo she had her scalp adorned with a henna tattoo.

'I was trying to adjust to going out and being bald and not feeling uncomfortable about it, and allowing myself to feel like it's normal to be bald.' She laughs as she tells me that children ran away in fright at the sight of her bald head; but, she says, 'Once I got the henna tattoo, it was a completely different feeling; it was easier to be bald. It was like wearing a hat or a scarf or a wig or something like that.'

She stands and turns, pulling up her shirt to show me the small sea turtle that she has had permanently inked onto the small of her back. This, too, was done in Byron Bay, shortly before her ovarian cancer had been diagnosed. She was depressed; work was stressful, and years of trying to fall pregnant through an IVF program were taking their toll. Her voice falters as she explains why she chose the turtle—as a symbol of longevity and fecundity. In light of her diagnosis with ovarian cancer, and the radical hysterectomy that was required as part of her treatment, not to mention the likely impact on her lifespan, her tattoo has assumed an unavoidable irony that she acknowledges, 'but no regret and still pleasure'.

For Toni, there is also a shamanic aspect of her tattoo that she hopes to be able to rechannel. 'Maybe it's clutching at straws, but perhaps that belief in longevity, and that hold on it for me, might have the impact of surviving.' As for the fertility aspect, there are other areas of her life, she hopes, where she can still be fertile: through her creativity and her garden.

'You won't stop at one,' she recalls the tattooist telling her, 'You'll be back for another.' And it's true that she is already planning a second—a bird in flight. She indicates where she already sees it in her mind's eye: soaring across the skin of her shoulder. She is a myotherapist, and she names the muscles that the tattoo will transverse. Both the turtle and the bird, it seems to me, are images of freedom, escape, transcendence. They both elude the pull of gravity: one in the buoyancy of the ocean, the other on the currents of the air. Her turtle tattoo is a symbol of hope, a small prayer offered up to the cosmos.

PETER BENDA has a caffeine molecule inked into his skin, just above the elbow joint of his left arm. The composition of the molecule is represented by a series of connecting lines and circles, and marks the milestone of the successful completion of his Masters degree. Peter is drawn to duality, and the tattoo is a memento of not simply his academic achievement but also of the cups of stimulant that he used as a crutch to get him through the rigorous period of learning and the long, gruelling hours of study.

'To me,' he says, 'it signifies this anxiety-ridden part of my character, and also a turning point in my life.'

At just over two metres tall, Peter is easy to spot in the café where I meet him, as it is frequented primarily by mothers and their toddlers. Despite his 36 years, by his own assessment, he dresses like a teenager — to which his black beanie and grey hoodie attest. His glasses and a small square of facial hair beneath his lower lip provide a slightly more aged counterpoint. Peter's tattoos aren't immediately apparent, and it is only the five piercings in his left ear that hint that body art may be a minor obsession of his.

Peter has five other tattoos apart from the caffeine molecule, (a smoking dinosaur, a small, turreted castle, and a large leg piece among them) and each marks 'an inflection point', as he calls it, in his life. They are reminders, keeping him aware of events that have shaped him and given his life texture.

A major rite of passage that hasn't yet made it onto the canvas of his skin is the birth of his daughter four years ago. 'Probably the most critical thing that's ever happened to me,' he says. He has ideas about the form that it will take, but cost is an issue, and he is not the sort to simply turn up at any studio and pick a design off the wall.

'I was thinking about some kind of sperm–egg motif, some chaotic kind of thing, but with a warmth to it.' There was, he believes, a driving biological need of his own to have a child, and his daughter brings him enormous joy. Both these factors compel him to mark the phenomenon of her existence onto his body.

For Peter, pain also has a role in the experience, and even in the decision to get a tattoo. 'I keep coming back to catharsis — to some notion of pain as release. Tattooing definitely has a power-over-the-flesh element to it.' It's also about staking a claim for him: 'This body is mine and no one else's.'

PETER PUTS ME onto Tim Dywelska, his tattooist of choice, who works out of a studio in an inner-city Melbourne suburb. The soft edges of Tim's Canadian accent and his neat, if adolescent, style of dress are somewhat at odds with his heavily tattooed arms. Dressed as he his in jeans and a T-shirt with sleeves down to his elbow, the tatts on his lower arms are the only ones that I can see. I like him. He's business-like, gallant, hip, and nerdy in quick succession, and sometimes all at once.

Tim's workplace is a far cry from Bev Nicholas' parlour in Williamstown. By comparison, Bev's seems a lonely little operation. The studio is humming at 11.00 a.m. on a weekday. Clients are seated in the waiting room, the phone is ringing, young tattooists are having 'consults' with clients in the line of booths, coffee orders are discussed, music is playing—it looks like a fun place to work.

Tim is brisk and efficient, apologising for making me wait while he talks to a client about a proposed tattoo. You can't just walk in here, pick a design, and half an hour later walk out with a tattoo. It's a custom studio; for the most part, clients bring in their ideas or maybe an image that they've sourced from somewhere. The tattooists, through discussions with their client, then draw the tattoo so it not only fits their client's taste, but also the site on their body that they've chosen to place it. They do a lot of large pieces here: whole arms, legs, and backs. So, with the time it takes to design and draw the individual tattoo, and because of the demand, it might be three months before you find yourself lying on a pristine bench feeling the repeated bee stings of the tattooing needle.

The wait has its upsides, however; because of the time and care taken, rarely do clients return dissatisfied. In fact, Tim says,

'I would have to say that never happens here, it takes too much preparation before we get to the actual tattooing.'

Tim is into Japanese-style tattoos—give him a large dragon, a samurai impaling himself, or a Koi fish to design and etch into your skin, and watch him smile. Americana is also close to his heart—'Your really clunky pin-up girls, your hearts, your little roses with the banners and the love hearts, sailing ships …'

This guy is hooked, it's obvious. It's not only his heavily tattooed arms that give it away, but the language he uses talking about what he does: the near reverence for the 'old-timers' who are his tattooing heroes, and the care that he takes preparing his tools as we talk.

I mention that I've been down to Williamstown to talk to Bev. 'I've heard she's lovely,' he says. 'If you can you believe it, I've never gone done to visit her. The grand old woman of Australian tattooing, and I've never gone to visit her.' He shakes his head in disbelief at his own neglect.

I try to get a bit philosophical, musing aloud about why so many people feel the need to mark their skin. Is it a way to connect with the body, or (echoing Peter Benda) is it something to do with the mind subduing the flesh?

'It just looks cool,' says Tim, but then he concedes, 'I say I get tattooed purely for the aesthetic reason, but it's a pretty powerful image to be walking into the pub and have both your arms tattooed.'

'Tattoos contain a lot of black,' he continues, 'not only for aesthetic reasons, but also to look heavy.'

When I ask if that was why he got tattooed, he hedges a little, 'Do I look like a tough guy to you?' (The answer is no.) He quotes 'a world-famous old-timer': 'It all boils down to guys want to look tough and girls want to look cute.'

Tim admits to having tattoos that represent heartbreak and emotional attachment, but also has others that he refers to as souvenirs; tattoos done by friends or by tattooists so famous that, when he found himself in their presence, he could not let the opportunity pass without getting them to leave their mark upon him.

This morning, Tim is tattooing 'a young lady', Emily, with an elaborate flamingo. It's a florid design, lush with flowers and a tropical sunset that's to be positioned on the back of her right calf. It will stretch from just above her heel, and end just below the back of her knee. He'll spend three or four hours on it today, by the end of which time she'll probably feel like she's been 'hit by a truck'. Completion of the tattoo will require at least two more sessions of a similar length.

'I'm actually heavily tattooed myself,' Tim says. 'I wouldn't do it to someone else if I didn't think I could get through it.' It 'pains' him, he says, to see tattooists who are not heavily tattooed themselves, because they 'obviously don't love this the way I do'. How could anyone tattoo someone with 'a big rib piece from here [he gestures to a point halfway down his leg] to here [he marks a place beneath his armpit] and not have gone through it yourself?' It's beyond him.

For the first time, seduced by Tim's enthusiasm for his art, I get the feeling that I am missing out on something. I can almost imagine commissioning Tim to tattoo me, except for the fact that I'm sure the modest vision of ink on skin that I entertain wouldn't have the scope to capture his imagination.

It's personal for Tim. For all his protestations that he is 'a commercial artist' who has to sell his skills to whoever is buying, he also admits that he is in an enviable position for a tattooist.

'I can pretty much pick and choose what I'd like to tattoo. But it comes down to, "Do I want to work today, or do I want to sit on my hands?" I can tell you, there's stuff I'd rather not do; but, in the end, I can do a good job, and that's what a person wants.'

Today, though, he is excited about the piece he will do for Emily. 'I really, really like this design. I know exactly how I'm going to do it. It's got this element of being quite complicated and busy, but when it comes down to it, it's quite loose and lovely. I'm really excited about it. You're really giving me, if I could be so bold,' he says to Emily, his gallantry coming to the fore, 'the chance to really do something artistic.'

I ask about the pain. 'There's no doubt that Emily here can feel it,' he says, as he works the needle across her skin, inking in the outline of the design.

'I'm a woman,' Emily says. 'I can handle it.'

Not long after Tim begins the outline of the flamingo, the young woman who has been taking calls at the front desk comes into Tim's booth. She's holding an image of a centaur that's been downloaded off the internet: all airbrushed, glossy flanks, rippled abs, and extravagant, wind-tossed hair, mane, and tail. A young man has just walked into the shop with it; he wants a tattoo based on the image. 'He's with his mum,' the receptionist says, half whispering, 'He's in detox, he's out today.'

It's not Tim's thing. 'Tell the kid he's going to have to wait and come in [to the shop] in a couple of months, and ask Owen if he'll do it. It's just not my forte.' He executes a flourish of the tattooing needle in the air over Emily's leg. 'Flamingos are my forte.'

The receptionist is reluctant to leave without having something more definite for the young man, but Tim is unmoved. 'Owen's our go-to-fantasy-portraiture kind of guy. I can do it, it's just not my trip.' He pauses for emphasis. 'It would *hurt* me to do it,' he says with finality, and then relents a little, suggesting the name of another studio that she could recommend to him.

After saying goodbye to Tim and Emily, I find myself on the footpath outside the shop with the young man out of detox for the day, and his mother. He looks sullen and twitchy; she looks like she wants to hug him, but she knows better and stands just outside the roiling current of his agitation. It's not the soothing touch of skin on skin that he craves, but the burn and sting of the tattoo needle.

I feel for both of them: for her, forced to babysit her adult son as he tries to wrestle his life back; and for him, struggling with addiction and having to go through this rite of passage with the indignity of having his mum beside him. His frustration at being thwarted in his quest is palpable. I can almost feel his ache for the stab of the needle; the desire to endure a pain that is only physical, and to have his skin indelibly marked with the image of a mythic figure symbolising virility and eminence in battle.

The lure of the tattoo; I'm beginning to understand it.

Alison Whyte's Boots: skinning and the human pelt

> Tan me hide when I'm dead, Fred,
> Tan me hide when I'm dead.
> So we tanned his hide when he died, Clyde,
> And that's it hanging on the shed.
> — 'Tie Me Kangaroo Down', Rolf Harris

THE BLADE IS CURVED and wickedly sharp. Every few strokes, the man hones it against a sharpening steel that hangs from a chain around his waist. He is sweating: the work is back-breaking—and intimate. He is close to the body that he's working on. Limbs are wrestled with, and their position adjusted. He holds the skin tightly as he pulls and stretches it, so the knife can separate it from the flesh: the flesh which it has, until now, protected and caressed more closely than a mother ever could.

The body moves with him as he slices, yielding its skin without a whimper. You would think that it would protest more at losing that which made it whole—without the skin it is but flesh, blood, and bone—but it is silent. There is no shrieking as the hide is stripped from the body to which it was cleaved. It is a carcass, after all; the man is not such a fiend that he would skin a horse alive.

At a certain point, further cutting is unnecessary, and it is more efficient to ease the skin off the carcass as if pulling a sock off a foot. A small tractor does the job; chains are attached to the skin and slowly, deliberately, the skin becomes a thing in and of itself, no longer an intrinsic part of the creature that once was. It lies with the other hides on the floor of the knackery yard, almost like a discarded coat.

ALISON WHYTE, an actress with red hair and creamy white skin prone to freckles, was being interviewed about her role in a television series with fellow actor Marcus Graham. When asked what she liked about her co-star, she admitted to coveting his 'beautiful, dark, silky skin—for a pair of boots'.

It's a response that might be judged bizarre by some, but perfectly understandable by others. Graham, a devastatingly handsome man, does possess beautiful skin. Boots made from his epidermis would be absolutely gorgeous. After all, most of us are content to wear the skin of animals on our feet or on our backs in the form of leather. We fashion it into clothing, luggage, book covers, watch bands, wallets, jewellery, hats, buttons, furniture, artworks, sports equipment, saddles, stock whips, bondage gear, musical instruments, and countless other objects.

I myself have coveted the soft suppleness of kid-leather gloves, desired the velvetiness of a suede wrap, and fantasised about owning a red leather couch. But the skin of our own kind, even the satiny hide of a handsome young man—that is a different matter.

OUR SKIN is integral to our identity. Without it, we would be unrecognisable. To live without skin is unthinkable. Literally and figuratively, our skin defines us. Only millimetres thick it may be, but it is our skin that is the wrapping on our corporeal package. It is what others see when they look at us; it is what is touched when we are touched by others. It is entangled and entwined with our sense of self. The body without skin is reduced to mere flesh and bone, identity torn from it, an unmitigated horror. We cannot see the person if we cannot see the skin. Having one's skin removed amounts to obliteration. As such, flaying has been employed as the most extreme form of torture or punishment, and instances of it exist in history as well as in myth and legend.

An early example of one who paid this penalty comes from Greek mythology. The satyr Marsyas found a flute that had been flung to earth in a fit of pique by the goddess Minerva. (Minerva had invented the flute, and played it well, but the goddess was vain. Cupid laughed at the contortions of her face as she played, and she threw it away.) Emboldened by the sweetness of the music that he drew from it, Marsyas challenged Apollo to a musical contest. Inevitably, Apollo won and, as punishment, flayed Marsyas alive for his impertinence. In Ovid's *Metamorphoses*, the satyr screams: 'Why tear me from myself?', as the god wreaks his terrible retribution. Marsyas' tormented cry reveals the intrinsic relationship between our skin and ourselves. With chilling detail, Ovid describes the horror of the torture:

> … the whole of him
> Was one huge wound, blood streaming everywhere,
> Sinews laid bare, veins naked, quivering
> And pulsing. You could count his twitching guts,
> And the tissues as the light shone through his ribs.

The Greek historian Herodotus, who lived in the fifth century BC, tells the story of another unfortunate who suffered the same savage punishment. Sisamnes, a Persian judge, accepted a bribe to deliver an unjust verdict, and was slain and flayed because of his transgression. The Persian emperor, Cambyses, decreed that Sisamnes' son, Otanes, should inherit his father's office. Sisamnes' skin was then used to upholster a chair where his son would sit when performing his duties. Presumably, his father's fate, daily impressed upon his own buttocks as he sat in judgement, ensured Otanes' own incorruptibility.

Herodotus' telling of the fate of Sisamnes is a mere aside within the much more grandiose project of recording and preserving the story of the Graeco–Persian wars, and the ultimate triumph of the tiny Greek states over the great Persian army. Herodotus was obviously diverted enough by the tale of Sisamenes' fate to interrupt his story of battles on land and sea, stunning victories, and doomed campaigns to relate it.

While the entire gory episode is dealt with in a handful of sentences, it was sufficiently vivid to fire the imagination of the 15th-century artist Gerard David, who plucked it from the tales of battles and stratagems to depict in 'The Flaying of the Corrupt Judge Sisamnes'.

The painter, who usually preferred religious subjects and themes, shows Sisamnes lying on a wooden table, his limbs tethered. Four men work on his body, slicing the skin and peeling it back. One holds a knife in his teeth, while with his hands he draws back the skin of Sisamnes' foot. Sisamnes himself stares straight ahead, as if he is merely undergoing an unpleasant, but hardly excruciating, medical procedure. His clothes lie under the table, his loins discreetly covered by a small white cloth. Perhaps he is drugged; you would hope so. Gathered around the head of the table is Cambyses, who prescribed the punishment, and other grandees of the court. In the background is a window into the future: Sisamnes' son Otanes, surrounded by other dignitaries, sits on the judge's chair that he has inherited, which is draped with his father's skin. Both groups stand in sombre, unexcitable tableaux; there is no hint of the sounds that would emanate from such an event, or the emotions that would accompany it: the shrieks and cries for mercy; the moist ripping sound as skin is torn from flesh; the curses and grunts from those carrying out the gruesome task. There is very little blood. Disgust, horror, empathy, malice — all are absent. It's a very orderly and placid affair in Gerard David's depiction.

Another flaying that inspired artists over the centuries was that of St Bartholomew, also known as Nathanael, in St John's gospel. Upon seeing Bartholomew/Nathanael, Jesus says, 'Behold an Israelite indeed, in whom there is no guile.'

Bartholomew, after Jesus' resurrection and ascension into heaven, went preaching in India and Armenia. It appears that he was too persuasive and guileless for his own good, achieving the conversion to Christianity of the King of Armenia's brother. The king was affronted by the apostle's success on the conversion front, and had him flayed alive and then beheaded. Some storytellers insist that he was also crucified upside down. The Catholics, never big on irony, made Bartholomew the patron saint of tanners and those who work with skins, including butchers, bookbinders, leatherworkers, and cobblers. Bartholomew's emblem is the very implement of his torture: a butcher's knife.

Michelangelo painted the saint in 'The Last Judgement' in the Sistine Chapel. Bartholomew is shown, knife in one hand, his own hide in the other, yet whole once more. The face on the flaccid pelt is apparently a self-portrait of the artist.

Even more grotesque, and (viewed from a 21st-century perspective) somewhat camp, is the artist Matteo di Giovanni's representation of the martyr. Here, Bartholomew stands, skinless, his flesh red-raw, his pelt draped around him like a fashionable wrap. As ever, the instrument of his torture is held in his right hand.

On 10 April 1661, centuries after St Bartholomew met his agonising end, Samuel Pepys recorded in his diary:

> Then to Rochester and there saw the Cathedrall ... Then away thence, observing the great doors of the church, which they say was covered with the skins of the Danes.

Rochester is the home of one of at least six churches in England whose doors are said to have been decorated with the hide of a marauding Dane during the reign of Aethelred I (830–71). Others include Hadstock, Copford, and Westminster Abbey. It was during this time that the Danes launched their main invasion of England. As explained by the Reverend George S. Tyack in his 1898 essay 'Human Skin on Church Doors', the English, while descended from Norsemen, were now Christians. Although they may have become 'tillers of the soil, and patient toilers in life's ways', confronted with the heinous offence of the sacking of a church, 'the fierce nature of the wild Viking showed itself':

> As the farmer nails the rook to his barn door for a warning to all the tribe, so the maddened English seized the poor wretch who had fallen into their hands, stripped from his

> quivering limbs his skin, and nailed it on the door of the church which he had sacrilegiously violated.

Here, at least, is some suggestion of the violence and single-mindedness required to carry out such a bloody and yet finicky task. Flaying someone alive requires more persistence than does mere decapitation, or hacking off a leg. A certain degree of finesse is required, not to mention the means to restrain your victim: ropes, perhaps, or strips of leather would do. Several sets of strong farmers' hands could serve just as well, as long as their owners had the stomach to stick to the task. For my part, the first sweep of the blade would have me backing off with the plea that the wretch had been punished enough, regardless of the heinousness of his crime.

In 2001, Alan Cooper of Oxford University, an 'ancient DNA expert' as identified by the BBC history website (just how old was he, one wonders), was allowed to retrieve a small fragment of leather from a larger sample kept at the Saffron Walden Museum in Essex. A label from 1883, attached to the original scrap of material, identified it as part of the hide of a sacrilegious Viking, lifted in 1791 from under the metal fittings of the Hadstock Cathedral door. Disappointingly, perhaps, the DNA that was extracted from the leather pointed to bovine rather than human origins for that particular relic of 'the skins of the Danes'.

The common theme in these accounts of flaying and death is an obvious one: violation of a law or tradition held to be sacred or immutable—impertinence towards a god or a king, corruption, or the sacking of a church—is punishable by obliteration, by the cleaving of the body from itself.

In England, this principle was enshrined in the judgements

bestowed on criminals who had drastically transgressed the laws of men. Prior to the mid-18th century, murderers were routinely sentenced to the gibbet after their execution.

Gibbeting—suspending a body in chains to be ravaged by the elements, birds, animals, and the natural processes of decay—was seen as a further punishment over and above mere death. A crime such as murder disqualified the perpetrator from the dignity of being buried in one piece.

Then, in 1752, in response to a push from surgeons eager to increase their knowledge and skills in anatomy and surgery, Parliament passed an Act that allowed judges to sentence murderers to dissection after their execution, rather than subjecting their bodies to the gibbet. The anatomist's slab denied criminals bodily integrity just as effectively as the gibbet. As a result of this legislation, more than one murderer's hide found itself flensed and tanned, and put to use as bookbinding.

We are animals, after all, and our skin can be preserved as leather just as easily as the pelt of any other creature. This ability to preserve human skin makes it possible to keep it as a relic, a talisman, or a trophy. Despite the very lack of life, a potency continues to inhabit human remains. The person who once animated the now-empty husk may have been obliterated, but something of their essence and power lives on—even if only within our minds.

Religious relics, and the more grisly aspects of some branches of witchcraft, are testament to this. Grimoires, books of magic or witchcraft, are said to be traditionally bound in human skin, and in Tanzania, until recently at least, a grisly trade in human skins operated. Used in rituals associated with witchcraft, the skins were transported to other countries, including South Africa,

Cameroon, Zambia, and the Democratic Republic of Congo.

In 2003, in an attempt to put a stop to the trade, the government of Tanzania exhibited flayed human skin at an international business fair to raise awareness of the practice.

The Aztecs practised the flaying of their captives in a ritual that demonstrated a more complicated view of the power of human remains, according to historian and writer Inga Clendinnen. In an essay published in 2002, 'Breaking the Mirror: from the Aztec feast of the flaying of men to organ transplantation', she draws on her earlier work, *Aztecs: an interpretation.* In it, she describes the events and rituals associated with Tlacaxipeualiztli, the Aztec 'feast of the flaying of men'. During this gory season of blood-letting and sacrifice, warriors slew their captives, offering their blood and bodies to the gods and other powers within their belief system. Once the victim was dead, he was flayed and, as Clendinnen describes it, his slayer donned the pelt and paraded around the streets for several days, continuing to wear it.

Clendinnen asserts that the wearing of his captive's skin, and other associated practices that followed, speaks of an intense identification between the warrior and this captive. While the warrior's family ate pieces of the victim's flesh, the warrior himself abstained. His family wailed and mourned as if it was their own young man who had died, fully aware that there was a good chance that he would one day meet the same fate.

Skin and other body parts have been taken as trophies of war in many battles and conflicts between cultures. This disturbing practice has also become a familiar aspect of accounts of serial killers and their crimes. Both in real-life cases and fictional stories of murderers who kill and kill again, the motif of the trophy is a powerful one. Such cases, when they come to light, draw our attention in powerful ways: they are sticky and cloying like blood, messy and repulsive, and hypnotic. The pull to immerse oneself in the visceral details is compelling, and can make one feel almost complicit in the crime through the voyeurism that it can incite. Two of the more unsettling and notorious examples of the 20th century are those of killers Eddie Gein and Jeffrey Dahmer.

In 1957, police stumbled upon the grisly obsession of Eddie Gein when searching his house in Wisconsin, USA. Gein was implicated in the disappearance of a local woman, but the police discovered almost-unimaginable horrors in addition to the headless carcass of the missing woman. As they searched his house, they discovered everyday objects, including lampshades, a wastebasket, and the upholstery on an armchair, that were fashioned from human skin. Even more bizarrely, the searchers came upon a belt made from nipples and, finally, a suit made entirely of human skin. These grisly discoveries have reverberated through popular culture ever since in movies and books, including *Psycho, The Texas Chainsaw Massacre*, and *Silence of the Lambs.*

In 1991, Jeffrey Dahmer was arrested following the discovery of body parts and photographs of dismembered corpses in his Milwaukee apartment. A young man, who would have been Dahmer's 18th victim, managed to escape, and waved down a police car to request help in removing the handcuffs that Dahmer had placed on him. On further investigation, the police found the grisly evidence of Dahmer's obsessions. Necrophilia, dismemberment, and cannibalism had all become part of his repertoire. The details of Dahmer's crimes and the macabre treatment of his victims' bodies sent reverberations of shock and dismay rippling through the public consciousness.

In his book, *The Shrine of Jeffrey Dahmer*, Brian Master recounts how Dahmer attempted to flay the corpse of one of his victims, about whom Dahmer said: '*him I like especially well*'. Another victim was also flayed, with more success this time, but Dahmer was unable to preserve the skin successfully, and had to eventually dispose of it. It seems that the skins of these unfortunate men, and the other body parts that Dahmer kept, were an attempt to retain some of the intimacy that he felt he had achieved with them, and which he was incapable of experiencing in a normal way.

Australia's recent history holds its own murderer who flayed her victim. On 29 February 2000, Katherine Knight, after having 'pleasurable' sexual intercourse with her de facto, John Price, stabbed him at least 37 times as he fled their bed and attempted to escape her frenzied attack, stumbling down the hall of his house in Aberdeen, New South Wales. Knight had worked for 14 years in abattoirs, and knew her way around the sharp edge of a knife. When Price was dead, she skinned him.

The judge who heard her case remarked on the expertise and steady hands required to remove Price's skin in one piece so that it formed a pelt. Knight then beheaded Price, hung his hide from a butcher's hook in a doorway, and cooked his head and pieces of his buttocks. She then arranged some of the cooked flesh with vegetables on plates intended for his children.

Price and Knight's relationship had been fiery, to say the least, with Knight having previously wounded her lover with a knife. In the days before his murder, Price had been to the police to take out an Apprehended Violence Order against her, and had told Knight that he wanted her out of his life.

In the opinion of the forensic psychiatrist who testified at her trial, the extreme way in which Knight mutilated Price's body indicated that the murder was motivated by revenge, and that killing Price gave her the opportunity to carry out the violent fantasies that she obviously harboured. Knight herself has never volunteered an explanation.

PERHAPS THE MOST INFAMOUS EXAMPLE of objects and artefacts manufactured from human skin comes from the Holocaust. After the Second World War, the sickening excesses of Karl Koch, the commandant of Buchenwald concentration camp, and his wife Ilsa were revealed. Feared by the inmates of the camp because of her brutality, Ilsa Koch became known as 'the Bitch of Buchenwald'. She reportedly collected human skin, and is reputed to have had a handbag made out of it.

According to interviews with prisoners of the camp, recorded in the *Buchenwald Report,* the Koch family and other Buchenwald SS officers had 'artworks' and objects such as lampshades in their homes made from human skin. Inmates claimed that Ilsa Koch would order the death of a prisoner who had an interesting or unusual tattoo with the express purpose of adding the skin to her collection.

Of all the atrocities and crimes committed against those imprisoned in Nazi concentration camps, the use of human skin as a material to be incorporated into and displayed as objects is one that is regularly held up as an emblem of that regime's evil.

Why, having ordered or witnessed the atrocities necessary for mass slaughter, would someone choose to own and display objects made from their victims' skin? Why did the Kochs' desire to have such things close to them and displayed in their homes? Did they regard them as curios; exotic trinkets that would impress their friends? Did they hope to normalise their actions by incorporating them into their everyday lives? Or did turning the remains of people into objects reinforce the idea for them that somehow these people—these Jews, these Gypsies, these homosexuals—were less than themselves? It was the ultimate act of objectification. In crafting these articles, they sought to turn those who were once alive into non-beings.

Yet flaying is not always enacted as a means of obliteration: it can be enacted with a view to commemoration, to fulfil a commercial contract, or as a selfless, but attention-grabbing, protest against exploitation.

In 2003, Australian newspapers reported on Carl Whittaker, a Mackay man who had had portions of his late father's tattooed skin tanned. While being treated for cancer, Whittaker's father had had several tattoos done, including one featuring a large eagle entwined with a snake, which covered much of his back. In his will, he requested that on his death the tattoos be removed and preserved. After his death, in 1999, his son, anxious to comply with his father's wishes, found a taxidermist to remove the tattooed skin, and a company that was willing to tan it. The tattoos were framed, and are displayed in Whittaker's home. Now a family heirloom, Whittaker hopes to pass them on to his daughter.

Tim Steiner, a young Swiss man, took a more commercial approach to the tattoo that he had inked onto his back by Wim Delvoye, a Belgian conceptual artist and tattooist. In 2008, the tattoo, which was of the Madonna with a death skull, was sold to a German collector for 150,000 euro in a deal that saw the artist, the gallery that represents him, and Steiner all get a cut. Steiner is contracted to display the tattoo three times a year and, on his death, the tattoo will be removed from his back, tanned, and passed on to the collector. As with any other artwork, it will be able to be sold, donated, or bequeathed as the owner sees fit.

For Ingrid Newkirk, animal-rights campaigner and leader of the People for the Ethical Treatment of Animals, the matter of the future disposal of her earthly remains is an opportunity to highlight the fate of many animals. In 2003, it was reported that Newkirk had left instructions in her will that, on her death, her flesh be barbecued, and her skin used to make leather products. She also suggested that her feet could be turned into umbrella stands. By proposing that her remains be subject to such treatment, she hoped to encourage people to rethink the way that most of us blithely exploit the creatures with which we share the planet—largely with scant regard for the suffering inflicted upon them.

No doubt Newkirk thought that she was suggesting something novel and shocking but, as we have seen, the utilisation of the human pelt has a long, if dishonourable, tradition.

Still, perhaps Alison Whyte can take comfort: with enough compliments about his dark, silky skin, and with Newkirk's example to follow, even Marcus Graham might be persuaded to part with a bit of himself on his death. What better testament to his beauty than a fine pair of boots made from his handsome epidermis, worn with jaunty aplomb by an actress with red hair? Or, better yet, for a man whose talents extend well beyond his handsomeness, to have a colourful account of his life and acting achievements recorded in a book bound (for posterity, of course) in his own epidermis. It wouldn't be the first time that such a volume has been produced.

Bound in Human Skin: anthropodermic bibliopegy

I find the application of this sort of leather to books a logical if *macabre* enterprise.
— Holbrook Jackson, *The Anatomy of Bibliomania*

IT IS LIKE OPENING a carefully wrapped gift, complete with the frisson of anticipation. First, the blue laminated card outlining the conditions of use of the library's special-collections material must be read. Now, the wrapped parcel can be removed from the plastic bag, and the white tape securing the orange cardboard sleeve loosened. Finally, but only now that I am wearing protective white-cotton gloves, the book is in my hands.

It is a plain but handsome volume with the title embossed on the spine in gold lettering, and a simple border pressed into the front and back covers. Finely grained and delicately textured, the colour of the leather binding is a dirty fawn flecked with spots of darker pigment. The pages are edged with gold, as are the margins of the leather where it has been folded over onto the inside of the cover. When I open the book, the first thing that I see is an inscription, underlined and in a neat flowing hand: *Bound in human skin.*

FROM THE MOMENT that I first read of the existence of books bound in human skin, it became something of a quest for me to find one in Australia. There are numerous references to such volumes held in many august institutions in the UK and the US. The Wellcome Library in London, an institution devoted to medicine, boasts a volume on virginity, pregnancy, and childbirth bound in a woman's skin. Published in 1663, the skin used for the current binding was tanned by a doctor who had acquired the text some 200 years later, and who had taken the skin from a corpse while a medical student. It seemed to him, he wrote in an inscription in the volume, that the book deserved a binding linked to its subject.

Brown University Library in Providence, Rhode Island, has a copy of Holbein's *Dance of Death*, published in 1816, bound in skin—a very apt pairing given the subject of Holbein's woodcuts, where death, represented by a skeleton, comes to call on rich and poor alike.

The Chronicles of Nawat Wuzeer Hyderabed, a book also believed to be bound in human skin, was found in the palace of the King of Delhi in 1857 during the Sepoy Mutiny, and is held at the Newberry Library, Chicago—although the only evidence of the binding's human origin (apart from the leather itself) is an unsigned inscription to that effect.

The Boston Athenaeum has a volume with the long and descriptive title, *Narrative of the life of James Allen: alias George Walton, alias Jonas Pierce, alias James H. York, alias Burley Grove, the highwayman: being his death-bed confession, to the warden of the Massachusetts state prison*. Allen requested that his memoir be bound in his own skin, and that the book be presented to one John Fenno, a plucky gentleman who had resisted Allen's attempt to rob him.

Westcountry Studies Library in Exeter is the home to a copy of Milton's poetical works bound in the skin of one George Cudmore. Cudmore was executed in 1832, but the book of poetry wasn't published until 1850. This fact leads inescapably to the realisation that the leather must have been kept rolled up somewhere until the opportunity to use it presented itself.

No doubt there are many other such volumes held in private collections around the globe. In April 2006, a ledger dating from the 1700s was found in a street in Leeds, presumably discarded after a burglary. Most of the bindings from skin-bound books were torn from the corpses of criminals killed by the state, from the bodies of those too poor to pay for their own burial, or perhaps those exhumed illegally by 'resurrectionists' who sold cadavers to surgeons eager to practise their skills. Notable exceptions do exist, of course, like the case of James Allen, the highwayman, mentioned above.

In his 1950 publication *The Anatomy of Bibliomania*, Holbrook Jackson, a leading bibliophile of his time, relates many and varied examples of anthropodermic bookbinding (some accounts have a hint of the apocryphal about them), several with the blessing of the original owner of the hide. These include a Russian poet who grasped the dubious opportunity of amputation to bind a book of his poems in the skin from his lopped-off leg, and a tubercular countess, charmingly complimented by a writer on her beautiful shoulders, who made arrangements that when she died her skin should be tanned and sent to him. Gallantly, he bound one of his own books in it. One can only applaud the pride, devotion, vanity, or pure cheek that moved such individuals to donate that most personal of relics—their skin—expressly for the purpose of encasing words.

A significant number of the books that I have mentioned date from the 18th century and the opening decades of the 19th century, the Age of Enlightenment, when reason and rational thought were championed and superstition was deplored. Phrenology, a method for determining a person's mental capabilities by the shape of their skull, and physiognomy, the analysis of a person's character based on their facial features, were developing as 'sciences', and the body was the site of much investigation. It was a time when indigenous peoples' remains were stolen and sold to museums and collectors for examination and display, and a hierarchy of the races was widely accepted. In such a cultural and intellectual environment as this, perhaps treating the hides of the socially inferior as curios kindled a titillating thrill rather than the disgust and outrage it would almost certainly produce today.

Conversely and paradoxically, the annihilation of the body after death was feared by many because of the widespread acceptance of the tenet of bodily resurrection; without a body, a person might be denied entry to heaven. This idea has held sway until as recently as the 1960s, when the Catholic Church finally relented and permitted cremation for its members.

As mentioned in the previous chapter, the importance placed on the integrity of the body at burial meant that allowing a murderer's body to be dissected after death was seen as increasing the severity of the sentence. On more than one occasion, convicted murderers had their skin tanned and used to sheath paper and ink just as closely as it once bound the flesh and blood that had animated it — sometimes around the very pages that told the story of their crime.

One such example is displayed in Moyse's Hall Museum in

Bury Saint Edmunds, Suffolk. It bears the following inscription on its frontispiece:

> The Binding of this book is the skin of the murderer William Corder taken from his body and tanned by myself in the year 1828.
> George Creed, Surgeon to the Suffolk Hospital

George Creed had dissected the body of Corder, the assailant in what became known as the Red Barn Murder. Corder had been convicted of murdering his lover, Maria Marten, and burying her body in a red barn. Creed had an account of Corder's trial bound in the leather made from his skin. Now a showpiece of the museum's collection, the book is displayed along with a bust made from Corder's death mask and a portion of his scalp with one ear attached.

My initial enquiries to two of Victoria's own august literary institutions, the State Library of Victoria and the University of Melbourne's Baillieu Library, asking if either had an example of anthropodermic bibliopegy in their collections both resulted in a faintly puzzled, 'No'.

I then sent a query to an online information service hosted by the national and state libraries, hoping that if I threw my net a little wider I might snag the rare creature that I was after. A few days later, an email came back from the National Library in Canberra. The curiosity of newly employed graduate at the library had been piqued by my query. In an intensive search of the catalogue, she had found an entry relating to a book with the unprepossessing title *The Poetical Works of Rogers, Campbell, J. Montgomery, Lamb, and Kirke White.*

In the notes of the book's entry was the information that I'd been hoping for: 'Pencilled note on front free end paper of EAP copy (in Dewey run) reads: "Bound in human skin." Confirmed; see file 430/01/00026.' Eureka! A skin-bound book was only a short flight with a discount airline away.

Before I could arrange my flight to Canberra, however, a call from the Special Collections curator at the University of Melbourne also got my hopes up. The Baillieu Library had received some items from the medical faculty's rare-book collection. One was an 18th-century Spanish anatomy text, and the leather binding … well, it just had a certain look about it.

My hopes were high. Published in 1757, *Anatomia Completa del Hombre* by Martin Martinez has an arcane look about it even if one approaches it with no suggestion that the binding may have come from the skin of a human. The leather binding has been folded over card rather than board, and has developed wrinkles and furrows, and in places has worn right through. It has a rosy tinge that gives the leather a look a little like pork crackling. The frontispiece is an engraving of a surgeon, knife in hand, about to dissect the cadaver on a table in front of him. Seated in a semicircle before the table is a coterie of bewigged observers. An ornate arch framing the proceedings is supported by a skeleton on one side and what looks to be a standing corpse on the other.

There was nothing to suggest, apart from the existence of other anthropodermically bound anatomy books from the same period, and its appearance, that it was indeed bound in human skin. The director of collections at the university was reluctant for the already-fragile volume to be subjected to any intrusive investigations of the binding, so for the moment I had to be content to leave it as a tantalising maybe.

What lies behind this enthralment with the notion of anthropodermic bibliopegy? To me, such volumes are the Hannibal Lecters of cultural objects, with the unholy allure of the serial killer swirling around them. There is the whiff of the occult about them, but also of the slaughterhouse. It is the twinning of the human hide (more graphic evidence of death and bodily obliteration would be hard to find) with that most potent symbol of human culture and learning—the book—that creates a dissonance at once fascinating and repulsive. And because of that dissonant pull, I flew to Canberra to hold *The Poetical Works of Rogers, Campbell, J. Montgomery, Lamb, and Kirke White* in my own prurient paws.

Enmeshed within and without this slim volume of poetry were enough contradictions to spin off a welter of thoughts. It represents the epitome of objectification of the body, entangled with the impulse to capture knowledge, culture, and experience in a form that will persevere long after the life that created it has been extinguished. An emphatic full stop to someone's life, yet entwined with the ongoing project of humanity, the volume also underlines our species' capacity for cruelty.

The potency that clings to all human remains conflicts with the sometimes-trite words and sentiments expressed in the verse that I found within its pages: 'On the sight of swans in Kensington Garden', 'To Jemima, Rose, and Elenore; three celebrated Scottish beauties', and 'Pastoral Song' are typical titles found in the book.

Finally, arching over any consideration of this or any anthropodermically bound book, is the thought of the labour and sheer gruesomeness of the process required to render the pelt that forms the cover. I have seen horses peeled like fruit in the knackers' yard, and it is an undertaking that requires skill, strength, and a certain hardness of heart.

As for the method employed to make human hide into useable leather, Holbrook Jackson offers an insight when he quotes Edwin Zaehnsdorf, who wrote in the trade journal *The Footwear Organiser* (and who also bound Brown University's *Dance of Death* in human hide) that 'manskin must be saturated for several days in a strong solution of alum, Roman vitriol, and common salt, dried in the shade, and dressed in ordinary fashion'. I have to confess to an unnatural affection for the term 'manskin'. It sounds distinctly Kipling-esque to me: something Shere Khan the tiger might mutter, licking his lips at the sight of a vulnerable child.

The edition of *The Poetical Works* held at the National Library was rebound in its current cover by a bookbinder, C. Egleton of Fleet Street, sometime after its publication. The small sticker bearing the binder's name on the inside-cover of the book makes no reference either to the origin of the binding or to who requested it to be done. It was not uncommon at the time for a book to be published with a temporary binding of board or card that would then be replaced with something more in keeping with the grandness or décor of the buyer's own library.

The book came into the library's collection in 1911, along with the rest of the material acquired from Edward Petherick. Petherick was a bookseller, book-collector, and bibliographer who assembled around 16,500 books, mainly relating to Australia, New Zealand, and the Pacific, with the intention that it would, in time, become part of an Australian national library. How this volume of second-tier British romantic poetry came to be bound in someone's pelt is not known. Petherick probably acquired it when he was living and working in London.

Obviously, questions concerning the provenance of human remains used in this way are serious ones. The sober and poignant reality is that the bindings of these books are the physical remains of what was once a person. Colourful tales of flamboyant individuals who consented to the use of the bodily parts notwithstanding, the majority of those whose skin went to bind books were the convicted, the vanquished, and the destitute. Yet, safe in the knowledge that a keen blade is unlikely to ever separate the hide from my own corpse, or from those belonging to those whom I love, it is possible to enjoy the macabre titillation excited by cradling a book that is bound in human skin.

Attitudes to how human remains are treated have changed dramatically over the last decades, although the terrain around the treatment of them is a shifting one. We quite happily view mummies plundered from Egyptian tombs, but no Australian museum would countenance the display of Indigenous remains in the same way. Indeed, Indigenous Australians have had mixed success in repatriating the remains of their ancestors housed in faraway museums and, in some cases, continue to demand their return. In the Victorian era, it was not uncommon for mourning bracelets to be made out of hair cut from the heads of the deceased. These were then worn in public displays of loss and grief. My own top drawer rattles happily with the baby teeth of my children, and curls of their infant hair nestle in zip-lock bags in boxes of mementos that mark the passing of their infanthood.

In 2007, in a case that has some parallels with claims for the return of Indigenous remains, the St Edmundsbury Borough Council, under the jurisdiction of which the Moyse's Hall Museum falls, considered a request from a descendent by marriage of William Corder to have his remains, including the book bound in his skin, turned over to her. Mrs Turner had been successful in having Corder's skeleton, held at London's Hunterian Museum, released, and it was subsequently cremated. The committee set up to review the matter unanimously voted to deny her request on the grounds that she was not Corder's direct descendent, and that closer relatives wished for the remains to be kept in situ at the museum.

BACK IN MELBOURNE, I ask Dr Shelley Robertson, senior pathologist at the Victorian Institute of Forensic Medicine (VIFM), to give an opinion as to what animal the leather that binds *Anatomia Completa del Hombre* was sourced from. As well as working at the VIFM, Shelley has her own consultancy business, is a published author whose work has been included in several true-crime anthologies, and has a special interest and expertise in aviation fatalities.

Even though I had already accepted that the chances of this 18th-century anatomy book being bound in human skin were remote, I have to confess that my heart does sink a little when Shelley's immediate impression is that the leather is not of human origin. She panders a little to the hope that I have left by telling me that it's very difficult to tell conclusively by simply looking at the leather, because the appearance will depend to a large degree on the methods and chemicals used to treat it, and to what extent it's been stretched. However, given the size and the patterning of the pores, her first guess would be pigskin. The leather is also largely devoid of obvious freckles and other small blemishes, possibly indicating that the skin had been protected from exposure to the sun by hair—another indication that it is animal rather than human skin.

DNA testing would be her first port of call in determining what the source of the leather is, she tells me; but, given the age and condition of the book, it would be a fairly destructive process. She can well understand why the book's custodians would be reluctant to let such an attempt proceed, especially given that there may not even be any DNA to be found.

'[To test for] DNA, you basically have to have tissue that's got a cell nucleus in it. Now, skin, and particularly the superficial layers of skin, are made up of the keratin protein … so there's no cells in the top layer of our skin and animal skin, anyway,' she says.

Any sample taken would need to be quite large to have a reasonable chance of finding a deeper section of the hide that included dermal cells where DNA could be found. And, even if DNA was successfully extracted, there's not much difference between pig and human skin even at the cellular level, Shelley tells me.

Pigskin is used regularly as a substitute in situations where ethics, practicality, or legal issues prevent the use of human tissue. Shelley has worked at the Police Forensic Centre in Melbourne, and tells me that when the police test-fire weapons to observe their effect on the human body, they often use pigskin stretched over a frame to simulate human skin.

And as for the difficulty of stripping the skin from a human cadaver: 'To skin a body, to put it crudely, is quite a simple dissection process,' Shelley says, 'and we [the VIFM] do it often. It's routinely part of a homicide autopsy, because it's important to document the injuries in a homicide, and a lot of injuries don't become manifest on the skin until some hours after they're sustained.'

In cases where a person has died immediately or soon after a beating, a subcutaneous dissection would be done in order to look for bruising in the tissue beneath the skin, because not enough time will have elapsed for those injuries to become visible on the skin. Anyone with basic butchery or surgical skills would find it quite a straightforward exercise.

Shelley and I both take photos of the book, and leave *Anatomia Completa del Hombre* to be returned to its place in the rare-books room. We are both a little disappointed.

THE VERY HELPFUL LIBRARIAN at the Newberry, when answering my questions about *The Chronicles of Nawat Wuzeer Hyderabed*, had also directed me to an online discussion of books bound in human skin. In addition to the more scholarly views on anthropodermic bookbinding, and those that expressed horror at such a practice, there were slightly more relaxed opinions on the subject. One was to the effect that nothing could be more sensible than to use a perfectly good natural resource such as the human hide as a book binding. Another respondent remarked on the satisfaction that one would get on handling a book bound in the skin from the buttocks of a loved one.

Perhaps that was the impetus behind (no pun intended) what is my favourite account of anthropodermic bibliopegy, revealed in Dard Hunter's autobiography, *My Life With Paper*, published in 1958. Skilled in almost every aspect of bookmaking—type design, illustration, papermaking, and calligraphy—Hunter relates how he received a commission from a young widow to produce a hand-lettered book of eulogies as a memorial to her late husband. When the time came to discuss the book's binding, Hunter went to the woman with a selection of materials from the bindery. She looked over the examples in a desultory manner, rejecting the pigskin, calf, and other fine materials before producing a roll of hide from her trunk. This delicately grained leather, she said, was what she wished the book to be bound in.

Hunter could not identify the animal from which the leather came, but was initially reluctant to admit his ignorance. Finally, he asked the widow to tell him the source of the material. It was, she told him, the tanned skin from her late husband's back.

Hunter ends the anecdote by recalling that, a few years later, he read that the woman was to be married again. He wondered if the second husband ever looked at the first husband's memorial book and 'thought of himself as Volume II'.

As Holbrook Jackson pointed out, there is a kind of logic to binding books in human hide given the 'close relationship between books and men', as he puts it. Binding a tome in the leather of a human—well, the relationship couldn't get much closer.

You're Nicked: the romance of fingerprints

> Every human being carries with him from his cradle to his grave certain physical marks which do not change their character, and by which he can always be identified …
> — *The Tragedy of Pudd'nhead Wilson*, Mark Twain, 1894

IN THE FOREWORD to George Wilton Wilton's 1938 book *Fingerprints: history, law and romance*, Dr Robert Heindl writes that the reforms adopted by 'civilized' countries in the late-19th century to 'the laws and methods of criminal procedure' did away with the 'robust methods of the Middle Ages'. 'Even the smallest of thumb screws was forbidden,' he laments, 'and as a result there was a painful lack of technical assistance.'

The *smallest* of thumbscrews? How churlish to deny intrepid crime fighters the smallest of anything. One does have to wonder if Dr Heindl was not, in fact, a bit of wag. The almost-jocular reference to torture, and his (surely) carefully chosen adjective of 'painful' to evoke a less-than-subtle pun, reveals a wry, black humour somewhat at odds with the romantic notion of fingerprints announced in the title of the book.

Fortunately, as Dr Heindl's foreword goes on to point out,

while those on the trail of criminals could no longer call on the rack or the ducking stool as a means of gathering evidence (they'd have to wait for the 21st century for the US to bring the latter back into the repertoire of the modern crime fighter), fingerprinting as a method of crime-scene analysis and identification was being developed.

In addition to giving the history of fingerprinting, the more pressing objective of Wilton's book, and one unannounced in the title, was to reassert the role of, and to gain recognition for, one Henry Faulds as the 'Pioneer of Modern Fingerprint Crime Detection'. Others had put themselves forward, or been crowned, as the one entitled to wear that particular honour, and there had been considerable jockeying for position in this emerging field for, as happenstance would have it, in the latter part of the 19th century, several people had begun to explore the possibility of using fingerprints as a means of identifying an individual.

As well as Faulds, they included Juan Vucetich, a Buenos Aires policeman, and Sir William Herschel, an employee of the East India Company. Others who had experimented or lobbied for the use of fingerprints as a crime-fighting technique were scientist and cousin of Charles Darwin, Francis Galton, and Sir Edward Henry, who was responsible for the adoption of the fingerprint system in India in the late 1890s, and then in 1901 by the British Government. In Australia, police fingerprint branches were established in New South Wales and Victoria in 1903.

Many centuries before these inquisitional and meticulous men had begun to explore the possibilities of identifying an individual from the pattern of the skin of their fingers and hand, however, the Chinese were using fingerprints to seal official documents, and Babylonian army deserters were forced to leave imprints of their thumbs and fingers as a permanent record of their guilt.

But back to Henry Faulds, who, in the mid- to late-19th century, was working as a missionary in Japan. Wilton describes Faulds's Eureka moment with regard to the uniqueness of fingerprints thus:

> Somewhere about 1878, while walking on the beach in the Bay of Yedo in Japan, Henry Faulds (1843–1930), then residing in Tokyo, found fragments of 'sun-baked' prehistoric pottery. They bore the finger-impressions of the Japanese potters left on the clay while still soft. Closer observation of these impressions during his residence in Japan, where he was located as the first Scottish medical missionary, led to his serious study of our finger-tip lineations or 'furrows' as he styled them, with all their possible bearings on questions of race and other matters. He came to the conclusion that our finger-patterns, so varied in design, arched, looped, whorled and mixed, as he described them, differed in every individual, irrespective of race and sex. In his striking phrase these patterns were 'forever-unchangeable,' …

Faulds published his observations on fingerprints, and his belief that they could be used in the detection of crime, in 1880 in the journal *Nature*. 'When bloody finger-marks or impressions on clay, glass, etc, exist,' he wrote, 'they may lead to the scientific identification of criminals.'

Some decades earlier on the sub-continent, Sir William Herschel, chief administrative officer of the East India Company, had begun experimenting with fingerprints as a way of ensuring his contractors honoured agreements with him, using them as an added measure to prevent an individual from 'repudiating his signature'. Later, as a magistrate, he proposed fingerprinting as a method of thwarting fraud. He also, according to a response that he wrote to Faulds's letter in *Nature*, which was published in the same journal, 'used it for pensioners whose vitality has been a distracting problem to Government in all countries.' You can almost hear him ask, 'Why do those old people insist on clinging to life?' He found it a useful method for identifying prisoners, too.

One of the first cases to be solved using fingerprints was in Argentina where, in 1892, two young brothers were found murdered. Their mother, Francesca Rojas, herself bearing wounds to her neck, accused a neighbour of the assault and the murder of her sons. The detective on the case found bloody fingerprints on a doorjamb in the house where the murder took place. The wood on which the prints were found was cut out and sent to La Plata Central Identification Bureau where Juan Vucetich was the director, along with finger impressions of Rojas and the accused man.

Vucetich, who had been developing a system for classifying fingerprints, matched the mother's prints with those on the doorjamb, and she confessed to the murder of her sons. Apparently, Francesca's lover was not as enamoured of her children as he was of her, and she cut their throats in order to be able to present herself to her paramour unencumbered. Vucetich's classification system is still used widely across South America today.

THE BELIEF that each of us is a unique individual, a collection of cells that has never existed before and will never be repeated again, is, in itself, a romantic idea. It is also a fact, and one for which the fingerprint has become something of an emblem. Formed in utero and, apart from scarring and growth, remaining, as Henry Faulds put it, 'forever-unchangeable' throughout our lives, our fingerprints are unique to each of us. While the pattern of fingerprints that is formed is largely genetic—and often results in siblings having a similar arrangement of features—not even identical twins, who share the same DNA, will have identical fingerprints.

Even that great satirist Mark Twain was not unmoved by the poetry of the notion of the absolute uniqueness of the individual, writing in *Pudd'nhead Wilson: a tale* that fingerprints were a person's 'physiological autograph'. 'This autograph,' he wrote, 'consists of the delicate lines or corrugations with which Nature marks the insides of the hands and the soles of the feet.'

Conan Doyle was also quick to take up the fictional possibilities that fingerprints offered. His Sherlock Holmes story *The Adventure of the Norwood Builder*, originally published in 1903, incorporated a bloody thumbprint in the plot. However, in this case, the great detective finds that the print left at the crime scene was deliberately planted in an attempt to incriminate a young man wrongly accused of murder.

Fingerprints are formed by patterns of ridges that are also found on the palms of our hands and on the soles and toes of our feet. These collections of minuscule crests and troughs are thought to have evolved to assist us with gripping objects and surfaces. The swoops and whorls of the ridges are the result of substructures called papillae that lie between the upper and lower layers of the epidermis. While the areas of the skin that have papillae have pores that exude sweat, they do not have oil glands, and so do not secrete sebum. Fingerprints will often contain sebum, however, because of the fingers touching other oily parts of the skin like the face or the hair. The characteristics of the ridges, including the number that occur and the patterns that they form, are what allow fingerprint experts to assign them to an individual.

Particular types of patterns are described in often geographical terms: delta, ridge, spur, lake, island, while the patterns formed are referred to in the more architectural expressions of arches, loops, and whorls.

ON A MILD, SUNNY WINTER AFTERNOON, I drive out to Forensic Drive in McLeod, a suburb in Melbourne's north, to meet Sergeant Steve Dunn of the Fingerprint Branch. In 1994, the branch was integrated, along with the State Forensic Science Laboratory, into the Victoria Forensic Science Centre (VFSC). I'm hoping to discover if the 'romance' of fingerprinting has been retained in the face of the adoption and refinement of other forensic tools that are used in crime detection.

In the face of relatively recent techniques such as DNA analysis, the examination of trace elements left by explosions, gunshots, and drugs, and high-tech surveillance, looking for fingerprints seems almost quaint. With the barrage of police dramas in recent years that rely heavily on the science of ballistics, autopsies, and blood-spatter analysis to hook their audience, one might think fingerprinting would be relegated to relic status. I'm happy to report that this is not the case.

The VFSC is a high-security site—'Secret Squirrel', as Steve Dunn puts it—and I must identify myself and be issued with a visitor's pass when I arrive. Steve, who has a 1970s feel about him, takes me on a tour of the branch's labs, set up in what appear to be portable buildings. He wouldn't be out of place in an episode of *Life on Mars*, the television show about a cop from the early-21st century who is transported back to the bad old days of the 1970s. Not that Steve is anything but thorough, skilled in his job, and above board; it's just something about the harlequin-patterned vest and black buttoned-up shirt.

He bombards me with technical information as he takes me through the various labs where fingerprints that are lifted from crime scenes are examined. If he can, he prefers to bring the print into the lab, he tells me, where he can bring the full range of technology to bear to get the best possible read. The first room that we enter is a light lab with a variety of lamps sporting lenses of different colours of the spectrum. What the fingerprint, or other mark, has been made with—blood, semen, oil, or a chemical of some sort—will determine what colour light will be most effective in showing it up.

The work that Steve does is all about looking for traces, the merest hint of things, and the first question that he asks himself when he's at a crime scene is, 'What can I see?'

Steve shows me a section of black plastic that he's been working on. It was taken from a car boot that had been completely lined with garbage bags and was found when the driver of the car had been pulled over for an unrelated matter. The fact of the plastic-lined boot was cause enough for suspicion that something sinister might be planned, or had already occurred, and Steve was called in to look for clues.

The marks on the plastic that are of interest to Steve have been coated with cyanoacrylate—SuperGlue, essentially. The Fingerprint Branch uses a version specially modified for their use, but Steve tells me that regular SuperGlue works just as well. Because cyanoacrylate is attracted to moisture, it will attach itself to latent fingerprints formed by sweat, for example, highlighting the traces left behind by a careless hand. This attraction to moisture, however, means it's not a suitable method for finding prints on paper or other porous material.

Steve turns out the lights in the lab, hands me a pair of goggles with orange lenses, and shines a blue light on the plastic.

Prints jump out at me. Well, to be truthful, they look more like indeterminate smudges, but Steve points out the fingerprints unhesitatingly.

Flicking on the light again, he shows me an enlarged photograph of a print which he took from a murder weapon that had been buried with a corpse, and another print that he lifted from a piece of masking tape taken from a gun and brought to light with SuperGlue. Light, he tells me, is a far more effective method for showing up fingerprints than the brush-and-powder method often used at the scene of a crime. The latter method is only effective for a single moisture reaction.

Incidentally, the person who dusts a crime scene with brush and powder in a finger-printing team is known as the 'dirty man', while the 'clean man' takes notes and photographs.

Other labs are devoted to humidifying cabinets and vacuum chambers within which objects can be placed to develop any latent prints left on them. Yet another lab is devoted to visualising and preserving latent prints with the chemical ninhydrin, which turns a purplish-blue when it comes into contact with amino acids. It's used commonly on paper, and Steve tells me that they used it much more in the past when credit-card transactions still required the processing of paper slips.

He assures me that all their technology is 'world's best practice', and tells me that they often liaise with luminaries of the crime-solving world like the FBI and Scotland Yard. It's getting less and less romantic all the time. I want fewer humidifiers and more dapper gents gently dusting powder over the handle of a dagger with a sable brush; but the tour goes on.

Steve hands me a stack of photographs of fingerprints, most of which look like ill-defined smears. One of the photos is part of the evidence assembled in an attempted-murder case. It shows a piece of glass with a print clearly visible in the blood. Steve explains how he's taken photos of the bloody mark using differently angled directional light, photoshopped the image, and added various stains to make the print clearer. He then compares it to a donor print—a print already on record—taken from the suspect in relation to a previous case. I wouldn't know where to start to find the points of similarity, but Steve confidently points out a ridge that he says is identical to both.

I ask him how he manages to convince a jury that what they're looking at has unquestionably been identified as belonging to one particular individual. In a trial situation, he says, the judge will usually outline the fingerprint-expert's qualifications and experience for the jury, and hopefully that will be enough to instil confidence in the specialist's expertise. Five years' training is required to do Steve's job, and he's been doing it for twenty-one. ('You give up your police career to be in Fingerprints,' Steve tells me. Not that he looks or sounds like he regrets it. He's been interested in fingerprints since finger-painting days, he says.) Nevertheless, it's quite common for the defence in a court case to bring in their own fingerprint expert to challenge the evidence.

Like most people, I suspect, I had assumed that, in this age of digital imaging and rapid computing, the final identification of a fingerprint would be done electronically. Not so. While the database where Steve works has between 2.3 and 2.8 million complete sets of prints, it can only throw up possible matches to a scanned-in image. It will select features of the print likely to be useful in determining a match, which the operator can accept or reject, or nominate others. These are called 'minutiae'. The computer will assign a score reflecting the likelihood that each is a match; but it is the fingerprint experts themselves who have to examine the images to positively identify that they came from the same finger. 'The computer is looking for an algorithm,' Steve says. 'We're looking for a print.'

'The first thing you look for,' Steve tells me, 'is a break in a continuous ridge.'

'Core' features and 'delta points' are also examined for similarity. 'Am I looking at a whorl? Where are my delta points?' Steve asks rhetorically, bringing up an image of a fingerprint on the screen.

Core features fall under four main groups: composite, arches, loops, and whorls, while delta points mark the places where the ridges change direction. The number of ridges between a central, or core, feature and a delta point is another identifying feature. Also, fingerprints generally slope towards the ulna bone, Steve tells me, which means that a reasonable guess can be make as to which hand—right or left—a print came from.

Once a positive identification is made, another three officers will check it to ensure that no mistakes have been made before it will be used in court as evidence.

If a person who has been fingerprinted is found innocent of a crime, their prints will be destroyed. For juveniles convicted of an offence, as long as they keep their nose clean for seven years their prints will also be destroyed. When I ask Steve how fingerprinting stands up to the other, perhaps sexier, forensic techniques of DNA analysis, ballistics, and toxicology, Steve points out that fingerprinting has the advantage of being quite inexpensive compared to other forensic services. 'We work hand in hand with our colleagues in biology,' he says, 'close and friendly.' For the record, the Victoria Fingerprint Branch positively identified 5200 people for the 12 months covering July 2008 to the end of June 2009.

Steve has shown me about all that he can, and the tour is over. He walks me to the security gate where I came in, and stops me as I'm about to go through the gate: 'I'll take your pass …'

'… and then I'm free to leave?' I finish for him.

'After we search you,' he says with an attempt at a roguish smile. Ah, the romance.

FINGERPRINTS ARE NOT undisputed territory, and there have been some spectacular and highly embarrassing incidents of fingerprints being linked to the wrong person.

In 2009, wrangling was still continuing in Scotland over former-detective Shirley McKie, who was accused of perjury when she denied that a thumbprint found on the doorframe of the bathroom at the home of a murder victim in 1997, and identified as hers, belonged to her. Detectives investigating the case insisted that it was hers and, as they had also identified another fingerprint at the scene as belonging to their favoured suspect, were apparently concerned that having to admit that their analysis of one fingerprint was wrong would throw doubt on the veracity of the other evidence.

Bitter and lengthy legal proceedings found McKie not guilty

of perjury, but not before she lost her job. She was eventually awarded €750,000 as compensation. Continuing public concern over the case resulted in an official inquiry into the affair that began in June 2009 and finally drew to a close in December 2009. Over the years, the five fingerprinting experts at the heart of the case were sacked.

Another well-documented case of fingerprint evidence being successfully challenged was in the case of the Madrid terrorist bombings in 2004, which killed 191 people. In an embarrassing stuff-up for the FBI, they wrongly matched a fingerprint found on a blue plastic bag near the scene of the bombing with a Portland attorney, Brandon Mayfield. Mayfield was a Muslim convert with an Egyptian wife. They held him in custody for two weeks before the Spanish police correctly identified the print as belonging to an Algerian. The reverberations of the incorrect match were felt throughout the US justice system, and prompted a review of 3000 cases relating to condemned prisoners.

According to reports published in the *New York Times*, the flawed identification came about because the workplace culture of the FBI discouraged its fingerprint experts from disagreeing with their superiors. In a report written by the then head of the quality-assurance unit of the FBI's laboratory division, it was revealed that there was insufficient checking on the first identification. The report, outlining the conclusions of the inquiry into the wrongful arrest of Mayfield, stated that human error, and not a 'methodology or a technology failure', was to blame.

Such snafus fail to shake the faith of Steve Dunn: while he concedes that there have been wrong identifications made using fingerprints, he believes that they do not represent a flaw in the theory, but rather errors in the likely matches that the computer throws up, or human error. And, he points out, no two individual fingerprints have ever been found to be the same. The FBI has over 300 million complete sets of fingerprints in their database, he tells me, and they are all unique.

SOME WEEKS AFTER my visit to Forensic Drive, I arrive at the courtroom in the Melbourne Magistrate's court where Steve is giving evidence in a committal hearing relating to a drugs case.

I'm running late, and Steve has already given his evidence and is now being cross-examined by the defence lawyer. Steve scrubs up well. He's had a haircut since I last saw him, and looks smart in a pale-yellow shirt, daffodil tie, and black suit.

Steve is explaining that he was the 'dirty man' on the job while his colleague was the 'clean man', which accounts for why most of the notes on the job sheet are not in Steve's handwriting.

Steve has lifted some prints from objects found in a drug kitchen, and the defence is attempting to cast doubt not only on the veracity of these particular prints, but also on the reliability of fingerprint evidence in general. I suspect there's no quicker way to get Steve's goat, and he is impatient and prickly under the lawyer's questioning. The fingerprint that Steve has lifted from a gas bottle is apparently one of the key pieces of evidence linking the defendant to the crime scene, and the lawyer is trying as hard as he can to undermine the worth of that smudge of grease.

'Wouldn't you expect to find more than one partial print on a shiny metal object?' the lawyer asks. Steve is exasperated: 'I see what I see. I find what I find.'

'Weren't there any other marks?' the lawyer asks.

There are always marks, Steve says, but 'fingerprints are the exception rather than the rule'.

The lawyer tries another tack: 'Fingerprints don't last very long, do they?'

Steve sighs deeply. 'It's a very difficult question to answer.' He then goes on to explain that there have been volumes of material written on the subject of the longevity of fingerprints, and much of it is contradictory. In addition, myriad factors—such as the surface on which the prints were made, the substance of the prints, and the atmospheric and weather conditions at the time that they were made—can all affect how long they will remain.

'You simply can't say when the fingerprint was placed on the bottle,' the lawyer goads.

'No,' Steve agrees, in a tone that suggests that the lawyer is merely reiterating what he's been saying all along.

The lawyer then asks how Steve can be certain that the print he lifted from the object is from the same person whose reference prints were on file. Now, it's Steve's expertise directly under attack. Mate, I want to tell the lawyer, Don't go there. But Steve is restrained, holding up the enlarged photo of the lifted print and jabbing the image with the hard point of his finger, then holding up a sheet of paper that is a printout of the defendant's prints on record and giving it a similarly savage stab. He is adamant: the print that he has lifted is a match with the right-hand little finger of the defendant.

The lawyer then alleges that Steve's identification, which has nominated ten points of comparison between the two prints, falls below the national standard.

'I marked ten points of comparison,' Steve tells him, 'but I found more.' Besides, he tells the lawyer, there is no fixed standard for how many points of comparison establish a reliable identification.

The lawyer asks Steve to concede that there have been issues of mistaken identity when fingerprint identification has relied on ten or fewer points of comparison. Steve rejects this assertion with a simple and emphatic 'No'.

It looks like this bit of argy-bargy is going to go on for some time, and I slip out of the courtroom, giving my respectful nod to the magistrate as I go. She doesn't notice. All her attention is on the exceedingly dull exchange going on. Steve's rising exasperation is the only thing occurring that is of interest.

I leave before it's decided if this particular individual has been implicated sufficiently by a trace of swirls, whorls, and ridges left by the careless touch of a hand. Only time will tell whether this particular physiological autograph, unique and forever unchangeable, will see the full force of the law brought down on his head by the expertise of Sergeant Steve Dunn.

Peering into the Abyss: burns injuries

> Burns are also a kind of an abyss to peer into, a brush with death, a facing of extremity.
>
> — Carter and Petro, *Rising from the Flames: the experience of the severely burnt*

As a child, I read a story, in a book dedicated to Catholic martyrs, about a young woman burnt at the stake for her piety. I can't remember which particular virgin saint she was; they have a tendency to blend into one—a young, devout, beautiful girl who is unreasonably called on to renounce her faith steadfastly refuses, is brutally tortured, possibly has her breasts cut off, is burnt at the stake, and then fast-tracked into paradise.

The author, obviously aware of the impressionability of her young readers, strayed from the objective facts of the case to reassure those pious children like myself still drinking in the fact that they may be called on to be tortured in order to be true to their faith: God took pity on the young woman, she wrote, and assumed her soul up into heaven before the pain became too much to bear.

It was a thin, possibly flammable straw, but I lunged at it.

Okay, if the heathens invaded and I was burnt at the stake for being a Catholic, God might well let my body fry like an egg—but with a minimum of pain, and my soul would be saved in the end. It wasn't much to hold on to, but it was better than nothing. I didn't question how the author knew such a thing, but surely it was obvious: how could God, who loved us so much, stand by and let a young girl burn? Burning was just about the worst thing that I could imagine in terms of pain and terror: the smell of my own smouldering skin and flesh, the sheer terror and heat of the flames, the choking smoke and depleting oxygen, the agony.

IN THE AGE following the genesis of the Earth, the Titans Prometheus and Epimetheus were charged with the project of creating man.

Epimetheus was the one who put in the hard yards, fashioning all the other animals, as well as giving shape to that pinnacle of creation, humanity. He gave each species different faculties to distinguish one from the other.

Prometheus, overseeing his brother's work, observed that Epimetheus had been so liberal with dishing out attributes that there remained nothing of worth that separated man from beast. Epimetheus, surveying his work, had to concede that this was indeed the case, and appealed to his brother for help.

Prometheus came up with a solution, persuading Minerva, the goddess of wisdom and warriors, to assist him in reaching the heavens. There, he approached the chariot of the sun as it travelled across the sky, and lit a torch from its fiery trail. He brought the flame back down to Earth, and presented the gift of fire to man, who promptly began to subdue all the animals and the rest of his environment to his desires.

But, as my mother always told me, fire is a good slave and a bad master. Every now and then, it throws off its shackles; and when it does, skin burns.

It is difficult to imagine how, without the ability to create and control fire, humans would have evolved to dominate the planet as we do now. In our relentless drive to shape our world—fashion implements, build structures and machines, cook food, and make ourselves clean and comfortable—fire and heat have so often been our tools. Yet we continue to have a love–hate relationship with fire. We are drawn to it, fascinated by the leap and glow of the flames of an open fire, which convey danger corralled as well as providing comfort and warmth.

The sheer spectacle of fire and its power to reduce everything in its path to ash makes it popular fodder for the media. The often-dramatic images that a fire produces—an explosion, a bushfire, or a building engulfed in a raging conflagration—stimulate an excitement that news producers are happy to exploit. For movie directors, the flaming figure staggering out of the burning building or the exploding vehicle has become a visual cliché.

Fire has also been seen in some instances as being cleansing; witches, heretics, and traitors were burnt to remove the taint of corruption, and to destroy the profane. Still, we fear the uncontrolled flame—and we are right to do so.

Burns are perhaps the most brutal assault that can be suffered by the skin, and the suffering doesn't end when the injury heals—nor does it, necessarily, grow less over time. The initial damage is just the beginning of a long ordeal. The aftermath of disfigurement, disability, and ongoing trauma can continue for the rest of a burns survivor's life.

A severe burn is an extreme event, a life-changing experience. To be burnt is to have some part of you irrevocably destroyed, to fall into a place that can never be fully escaped from.

How do those who stand on the edge of the abyss of a severe burns injury, and those who have plunged into it, resist the pull of vertigo to fall further?

IN THE WAKE of the February 2009 Black Saturday bushfires that ravaged large parts of the Australian state of Victoria, there was ample opportunity for those who were not in the path of the flames to contemplate what it must have been like to face them.

In the weeks immediately following the fires, and then again while the Royal Commission into the bushfires was being held, tale after terrifying tale was told—of narrow escapes, of bad decisions that resulted in tragedy, of the grief and horror of telephone calls that relayed the last moments of loved ones trapped by the unstoppable blaze.

More than 170 people died in the fires, or as a result of the injuries that they received. Hundreds were injured, more than 70 severely so. Dozens were admitted to hospital, with 20 of the most seriously burnt admitted to the Alfred Hospital's burns unit.

In one telling sentence from a research article examining the immediate impact of the fires, published in the *Medical Journal of Australia*, the bald reality of the consequences of that day are revealed: 'Most bushfire victims either died, or survived with minor injuries.' In other words, there was little middle ground: those unlucky enough to come into contact with the flames were either killed or barely scorched.

THE ALFRED HOSPITAL in Melbourne is home to the Victorian Adult Burns Service. It is the state-wide provider of care for adults with major burns injuries, including reconstructive surgery for burns patients, rehabilitation services, and an outpatients' clinic. In addition to dealing with the everyday burns emergencies resulting from house fires, industrial accidents, car crashes, and those who set fire to themselves, it, along with hospitals around the country, treated survivors of the 2002 Bali bombings, and was in the frontline response to the Black Saturday bushfires.

Days before the fire, with forecasts of unprecedented heat and searing winds, the state government and the emergency services were warning all Victorians that Saturday 7 February was going to be a shocker. Early in the morning of that terrible day, and in anticipation of what was to come, the State Health Emergency Response Plan was activated. Ambulance services and hospitals across the state were on alert and prepared for the worst. Patients at the Alfred were discharged or moved to make room for the burns victims, whom it was feared would come in a deluge. Preparations were made to increase the numbers of ICU beds available, and burns units assembled their teams.

Yvonne Singer is the Victorian State Burns Education Program Coordinator in the Victorian Adult Burns Service based at the Alfred Hospital. Her role is to examine and develop guidelines for the state's disaster planning. Appointed to the position in mid-2009, she has worked in the Alfred Burns Unit as a nurse, and was the care coordinator for burns patients at the Alfred for nine years.

Yvonne was working in the unit when the Bali bombings occurred in 2002. To my Irish–Anglo eyes, she is exotically pretty, with black hair and dark eyes. More importantly, she has extensive experience working with burns patients, and a formidable knowledge not only about the injuries caused by burns but also their care and treatment. She has warmth and a confidence in her skills, backed up by years of experience, intelligence, and compassion, and strikes me as someone who would be of immense comfort to anyone who is hurt, vulnerable, and frightened.

When Heather Cleland, the director of the Alfred's Burns Service, called her on the night of 7 February 2009, Yvonne was still in the role of care coordinator at the burns unit. 'I think you'd better come in to the hospital,' Cleland told her.

Yvonne doesn't normally smoke, but something about the knowledge of what the next hours and days would bring made her want a cigarette. Curious, isn't it? A woman about to treat the victims of the worst bushfire disaster that the country has ever seen strikes a match and lifts the small flame to a tube of incendiary material, and inhales the hot, acrid smoke.

Yvonne was at the hospital for the next 24 hours. 'Even after that,' she says, 'I couldn't sleep at all.'

THE SKIN IS SUSCEPTIBLE to burning not just from heat but also from chemicals, electricity, radiation, and from the sun.

Burns inflict severe damage to blood vessels, disrupting cell walls, and causing fluids to seep out, which results in blisters and swelling. The body generates a largely uncontrolled inflammatory reaction in response to the amount of debris created by the destruction of tissue, which needs to be removed and leads to scarring. It then becomes a circular effect, with the inflammation contributing to the scar, and the scar provoking more inflammation.

For medical-classification purposes, burns are categorised

as being either first, second, or third degree, depending on the severity of the damage to the skin and the underlying tissue. In a first-degree burn, damage has been restricted to the outer layer of the skin, or epidermis. A second-degree burn will result in damage both to the epidermis and the layer of skin beneath it, the dermis. Both first- and second-degree burns will usually heal without skin grafts, as sufficient skin tissue remains in the wound bed to rebuild the skin.

However, third-degree burns will have damaged or completely destroyed tissue right through to the fullest depth of the skin, and possibly the underlying tissue as well. Skin grafts will be required and, often, artificial materials will need to be laid over the exposed tissue to protect it, and to stimulate new skin growth.

Skin substitutes have been developed that can provide a temporary cover while the wound from a burn heals beneath. The application of these materials also helps to reduce pain by covering exposed nerve-endings. These skin substitutes may contain biological tissue such as pig collagen or shark cartilage. Allografts (tissue from another individual's body) and autografts (skin grafts taken from non-injured sites on the patient's body) are also used as coverings.

The Alfred Hospital has also recently reopened its tissue laboratory, where cultured epidermal autografts are grown from a small amount of a patient's skin, which is taken in a procedure called a 'punch biopsy'. This method of producing a graft is saved for patients with a massive burns injury where donor sites for regular grafts are particularly limited.

The skin cells taken in the biopsy are put into a solution in Petri dishes where they expand and multiply until they can be reapplied to the patient's body. These small sheets of skin (around ten square centimetres) are not as robust as regular skin grafts, and may contain small gaps and holes in the tissue.

Before being applied to the patient's wound, the sheet of skin is put through a contraption that stretches it to the point where it begins to look like the elasticised mesh wrapping that you might see around a boned roast at your butcher, or a fishnet stocking. The meshed skin is then laid on the wound, and as the skin cells begin to regenerate in the wound bed they fill in the interstices of the graft. This method of grafting will result in scarring that has a crosshatch appearance on the skin.

A measure called the '9 per cent rule' is used to estimate the amount of the total body-surface area (TBSA) that has been injured as a result of a burn. Imagine a cut-out paper doll: the front and back of the upper torso are accorded 9 per cent of TBSA each, as are the front and back of the lower torso. The front and back of each leg also account for 9 per cent, while the whole of each arm is the same. The face represents 4.5 per cent of TBSA, the genitals 1 per cent, and the combined area of the palm and fingers of the hand 1 per cent. Burns to over 10 per cent of the body are considered serious enough to warrant admission to a specialist burns facility and, in Victoria, a severe burn is defined as one that affects 20 per cent or more of TBSA.

The treatment of a severe burn is drastic — savage, even — but necessary if the patient is to be given the greatest chance of survival. In order to give a burn wound the best opportunity to heal, medical staff concentrate on salvaging as much tissue as possible. With a burn injury, there will always be a zone of necrosis or dead tissue. This is will be removed as soon as possible to promote healing. The already-damaged skin and tissue is further assaulted with blades to cut away dead and dying tissue in a surgical procedure called 'debridement'. Around the tissue beyond repair will be an area of damaged, but still viable, tissue—a zone of stasis—and this is where most of the efforts of a medical burns team will be concentrated. If the tissue in that zone of stasis is lost, a partial-thickness burn can turn into a full-thickness, or third-degree, burn.

An escharotomy may need to be performed: leathery, dead tissue is surgically incised to allow the swelling tissue beneath to expand without constricting blood flow or, when the burn occurs around the chest, breathing.

Fasciotomy, cutting through connective tissue to relieve pressure caused by inflammation and fluid build-up, may also be required.

Initial burn treatment is all geared towards early wound-closure. A burn that heals within ten days will result in little, if any, scarring. If that time period blows out to 21 days, there's about an 80 per cent chance of raised or hypertrophic scarring.

Yvonne shows me a series of photographs taken over six days of a young man whom she treated who had received partial-thickness burns to his face. In the first photograph, his skin is ravaged, swollen, and bloody. A skin substitute called Biobrane was placed over his wounds to aid healing, and Yvonne points out the texture of the dressing where it has adhered to his face, as it is designed to do. Over the course of a few days, the speed of healing is dramatic, and it is marvellous and gratifying to see his face emerge from the damaged, bloodied, and puffy flesh evident on the day of his injury.

Yvonne smiles at the result of her handiwork. 'It's like unwrapping a present,' she says, describing how she felt each day removing the dressing to see the healing that had taken place on her patient's skin.

YOUNG ADULT MALES make up about 70 per cent of burns patients across the country.

In the case of a severe burn, a patient may have to be in hospital for up six months with another six months in rehabilitation. When they finally do return home, it might be two years before they get back to work, if they do at all. As Yvonne points out, all that time adds to a long period of being out of their normal society and activities, particularly for young men in the prime of their lives. If you're someone who is at a point in your life at which you're 'defining yourself and wanting to become a man', she says, 'it's a devastating injury'.

Cooking meat. It is an image Yvonne refers to more than once when describing the effects of heat on flesh, particularly if first aid isn't applied immediately.

She has frequently seen young men turn up at the burns unit the morning after a big night out. Perhaps they began by just sitting around a campfire with their mates. A few drinks later, they might be jumping over the flames, playing at fire-eating, or even giving themselves the cheap thrill of watching butane cans explode. A stumble, a misjudgement, and suddenly their clothes are on fire. Often, if alcohol is involved, they, and those watching, may be slow to respond, or they may even run and fuel the flames, forgetting the mantra of 'stop, drop, and roll', which is taught to every school child. Then, when the fire is out, if cool running water isn't immediately applied to the burn, it's like leaving roasted meat to rest: the cooking process continues. It may not be until the next morning, when they wake up to a hangover and a painful injury, that they realise the extent of the burn.

A severe burns injury, as opposed to other types of trauma, Yvonne says, is unique in the way the patient experiences pain. With other injuries — a broken limb, for example — the pain is most severe at the moment of injury. Treatment and medication will mean that pain is experienced as 'a one way slide down a scale of intense to less', as Yvonne puts it. For burns patients, though, she says, as her hands trace the trajectory of a roller-coaster, 'it just goes up and down and around and around'.

The good news is that the management of pain has improved radically over the last few decades. In Yvonne's time working with burns injuries, she says, the management of pain has become an area of expertise in and of itself, and a burgeoning area of research. She recalls that, early in her career, simply changing a patient's dressing could be a 'white-knuckle ride' of pain, imperfectly smoothed out with intravenous morphine. Now, morphine is not the only option: they regularly call on the services of an anaesthetist and an anaesthetic nurse for patients requiring big procedures.

Yvonne describes a patient who had received burns to 95 per cent of his body — 'one of our greatest tales of survival' — whose scarring had left large areas of his skin extremely tight, particularly in his underarm area, making the simplest of movements painful. Further limiting his movement was his fear and anxiety.

In his physiotherapy sessions, he was given ketamine, an analgesic that also acts as a disinhibitor, which makes it a popular recreational drug. He was given a visualisation exercise where he pretended that he was under water. 'Swim to the top,' Yvonne urged, and she, the physio, and the occupational therapist watched in amazement as he raised his arms above his head, making strokes as if he were propelling himself through water. In normal circumstances, this range of movement was completely beyond him. She smiles broadly at the memory of the moment.

'The relationships you form with patients must be quite intense,' I offer.

'They just trust you implicitly,' Yvonne says of her patients, 'It's …' she pauses. 'I'm looking for a word that's the opposite of burden,' she says. In the end, she settles for responsibility, but I sense that's not quite adequate for what she's trying to express. Blessing, perhaps—or maybe simply a gift.

Yvonne says she and the other staff in the burns unit are often humbled by the strength and courage that their patients exhibit. It's not just the way that they meet the physical challenges of dealing with the pain and discomfort, she says, but their adjustment to a sometimes drastically changed body image brought about by scarring and disfigurement: 'The dignity and grace with which they [carry themselves], and the way they want to keep going; it's taught us a lot of life lessons. It certainly has for me.'

NOT SURPRISINGLY, Yvonne says that she has a fear of being burnt, and admits to being scared of flames. She is methodical about turning appliances off before leaving the house, and would never even duck out with the washing machine still on. Social events that involve alcohol and flames, like barbeques, make her wary.

As she speaks, memories of my own carelessness and the situations that could have turned nasty, but thankfully never did, surface: that New Year's Eve where flares were set off; parties around bonfires with drunk young men demonstrating their bravado; a crowded nightclub when a young female fire-eater set her voluminous, highly flammable dress alight. The thoughts make me shudder.

I have already spent considerable time looking at pictures of burns injuries before Yvonne shows me more. It is not easy. The images provoke a visceral, stomach-churning response. The swelling, the sloughing of tissue, the charred skin, black and flaking, the intense redness, the distortion of limbs, the fluid gathering in blisters; it all screams pain. Of course, if you suffer third-degree burns, because the nerves are destroyed along with the skin the injury is not initially very painful. But the disfigurement will be substantial.

I struggle with the thought of how I would deal with the reality of severe scarring, particularly to my face, and ask Yvonne how her patients cope.

'My experience is that they're not really thinking about the scarring right at the start of the injury,' she says. 'They're either just thankful to be alive, and their wounds and their burns are still going through processes; it's not the end-point of scarring yet. They may or may not have realistic expectations of what they're going to look like at the end …' Scarring, Yvonne says, begins once the wound is closed, and will continue to develop for the next 18 months.

Yvonne also has on her mind the mental wellbeing of those who find the severely burnt. Often, they are people from the emergency services: the ambos, the coppers, and the smokies, who have seen horrific things before in their line of work. Sometimes they are family members, neighbours, a Country Fire Authority volunteer on their first outing, or strangers who happen upon them. 'How do they deal with seeing someone on fire?' she wonders. 'It would just be such a horrible experience.'

The treatment of burns has also come a long way in the last ten years or so: the rate of healing has increased and the amount of scarring has reduced. Early excision of the burn to remove all the dead tissue is now standard practice.

In the early 1990s, Yvonne tells me, medical staff might wait for five days to assess the depth of a burns wound before beginning to debride the dead tissue. Now, she says, it's acknowledged that the chances of surviving a major burns injury are much better if all dead tissue is excised within 24 hours.

Adequate fluid resuscitation is crucial for a burns patient, as they lose large amounts of fluid through their wounds. In response to trauma, the body naturally preserves blood flow to the vital organs, which can result in peripheral shutdown to the skin, meaning blood flow is severely restricted to the capillaries. To avoid this, the patient must be kept warm and hydrated and perfused with an adequate blood volume to ensure good circulation.

She tells me of marathon theatre sessions, where the surgeons, fully gowned, operate for hours in theatres air-conditioned to 32 degrees and where the humidity is deliberately kept at high levels in order to give the patients, who are losing heat and moisture out of their wounds, the greatest chance of survival.

WHEN KAREN MORAN first saw Grazi Lisciotto, she recalls thinking, 'Shit, what the hell happened to him?'

Karen was attending her first retreat offered by Burn Foundation Australia (BFA) and, although a burn survivor herself for 40 years, had never seen another severely scarred person in the flesh. As she describes her reaction to his ruined face, seamed scalp, and heavily scarred arms, Grazi, sitting opposite her at the table at which we're having tea and cake, begins to chuckle. By the time she's finished her story, he is laughing raucously, and says to my dismayed expression, 'Don't worry, I get that.'

To some degree, sending a request to the BFA to interview a burns survivor makes me feel like an ambulance chaser. Still, I can't adequately explore the subject of burns injuries without speaking to someone who has suffered that fate. There is almost a mythic horror surrounding burns: a familiar narrative in popular culture is that those who are burnt become monsters, impossible to love because of their disfigurement. And, frankly, the prospect of meeting someone who has been severely burnt is daunting. I will be confronting one of my worst fears, and hoping to discover if I have the wit and the compassion to see through my fear to the person that has emerged on the other side.

I needn't have worried. Karen and Grazi are, respectively, the CEO and vice president of BFA. Relatively new recruits to their positions, they are in Melbourne to raise their organisation's profile with Victorian hospitals and the state government. They plan to run the BFA's annual retreat in Victoria in 2010 and, in the wake of the bushfires in February, are hoping for some government support. There is a warm camaraderie between them, and a shared passion for minimising the pain and grief of other burns survivors. Frank and open, they are, to my tentative ears, searingly honest with each other and about themselves, quick to laugh, and passionate about their work for the BFA.

Grazi suffered burns to 67 per cent of his body. To put that into perspective, the only skin that wasn't burnt was that covered by the cotton singlet and underpants which he was wearing at the time. His mouth, as he describes it, 'was a mess', and the reconstruction of it gives him a look a little like that of Homer Simpson.

'They took this bit,' he says, gesturing to the area beneath his nose, 'from here', he indicates an area at the back of his head. He is completely bald, his scalp shiny and marked with the scars of the grafts as well as the burn. The tips of his ears are gone, and his eyebrows are small tufts of hair. Both arms are marked by the cross-hatch of mesh grafts and, at one point, as he goes through the surgery that put his skin back together, he holds up his hand to show me the small area of his right palm not touched by the flames.

The main thing that I remember, though, when I recall his appearance, are his bright brown eyes.

The clichés present themselves unbidden: the phoenix rising from the ashes, a baptism of fire. But Grazi's story can't be told without evoking them, so I might as well announce them up front. Grazi set fire to himself. It's a hard fact, and one that he doesn't shy away from. Not now. But that wasn't always the case.

Grazi has been a survivor for three years; that's in his estimation. He was burnt six years ago, and for the first three years, he says, 'I was alive, but dead.' He attended his rehabilitation and medical appointments, and performed the everyday tasks required of him; but, as he says, 'There was nothing inside.'

It was discovering the BFA and meeting other survivors like Karen that brought him back to life. 'I saw these people that had injuries, or were burnt 20, 30, 40 years ago, that were married, with kids. It was like, "Hey, if they can do that, why can't I?"'

The day that he returned home from his first BFA retreat, his family could immediately see the difference. 'It was a man that came home,' he says. He began volunteering at the BFA, and then, in 2008, was asked to go on the board.

Before his burn, Grazi had suffered from depression for years, and was someone who, in his own eyes, had failed as a husband and as a father. These are harsh judgements, and ones that Grazi offers up himself. It seems to me that he proffers them as a point of self-respect, or perhaps because of a pact that he has made with himself that, as an advocate and a mentor for those who suffer from the pain, disfigurement and disability of burning, he will not shy away from talking about his own experience.

'I'm a better person than I was, I know,' he says of himself now. 'The thing is, I'm always thinking, "Jeez, you had to learn the hard way."'

He laughs heartily and then, more soberly, says, 'It's a big pill to swallow. It's not easy,' and, as if acknowledging to himself, what he's come through and what he's achieved, he repeats, 'It's not easy.' He feels a debt to those who have helped him overcome the despair and the physical assault on his body. 'I'm not going to waste the six years and all the help that all the people have done for me. I'd say it was a waste if I went down that road.'

Grazi was out of hospital and rehab in almost record time after his burn. He threw himself into recovery partly because he saw it as his punishment for trying to kill himself — the pain that he had caused his family was evident on their stricken faces when he came to consciousness in the hospital. 'You do the crime, you do the time,' he says.

He no longer suffers from depression. 'That's the disappointing bit,' he says, that he couldn't overcome the black dog before his injury. It's still a bit of a mystery to him.

Since becoming involved in the BFA and hearing other survivors' stories, he's learned that this is not an uncommon experience: going through the suffering of being burnt has given them a new perspective on life, made them happy to be alive. 'It's the burn that changes you,' he says. 'I don't know why.'

'I wouldn't say, "Take my scars away,"' he tells me. And I believe him. However, he would change one thing at least. 'If I could swap my scars here,' he lifts both hands to his face, almost cupping his head in the bowl of his palms, 'and put them on my chest, I would.'

There are a lot of less dramatic and painful ways to end your life than setting fire to yourself, and I have to ask Grazi, 'Why choose burning?'

'I can't answer that question,' he says, and then, as if he has silently repeated the question to himself, 'I can't answer that question. I don't know.'

Trying to come up with an explanation used to consume him, but three years ago, when he found the BFA, he said to himself, 'No, stop it, mate. That's it.' Now, he thinks of himself as lucky.

'You're very lucky,' says Karen, who has been listening to Grazi tell his story. She knows it well, has heard it before. I get the impression that these two have shared more than one dark night of the soul.

'Because if I hanged myself,' Grazi continues, 'I wouldn't be here. So, you know what, I'm glad. I'm glad I did what I did.'

Grazi is an inspiration to others that have been burnt, Karen says to me, and has helped them embrace life again the way that he has. She eyeballs Grazi across the table. 'And it's sad that it happened to you.' But the fire that left him permanently scarred has re-made him into a man who is a role model for other burn survivors, she tells me.

When Grazi first became involved with BFA, if someone asked him how he was burnt, he would answer, 'House fire.' It was partly true—the house did burn—but he would be silently praying that they didn't ask how the fire started. 'The fear was bad,' he says.

Now, if people ask, he tells them the whole truth. 'And it's good,' he says.

'They just love you for what you are,' Karen tells Grazi, of her family and friends who have met him. 'You're burnt. Who cares? It doesn't matter how it happened. You should give yourself a tap on the back for that, because that's a hard thing. I burnt myself, too. But I was a little kid; it was different.'

Grazi used to imagine that he was the only person who had set fire to themselves and survived, until a couple of years ago when, through a BFA retreat, he met another man who had tried to end his life the same way. 'When I met another one, I was rapt.' He says this so enthusiastically, almost with pleasure, that both Karen and I guffaw. 'When I told him that I'd done it [too], he started bawling his eyes out. He clung to me that whole retreat.'

Being severely burnt has also meant a complete life change for Grazi. While Karen has a full-time job as well as fulfilling her CEO role with BFA, Grazi was unable to return to his pre-burns construction-industry job. His role as vice president, and the advocacy, counselling, and support that he offers other burns survivors, is fulltime.

He has had to learn a whole new set of skills that are vastly different from those that he used in the construction industry. Running a charity has been a crash-course in fundraising, government and fiduciary regulations, and in the ramifications of policy. All the work that he does for the BFA is on a volunteer basis, and the organisation is financed through donations and limited government funding. It's a battle to find the money to keep the organisation afloat.

Grazi is also at Concord Hospital in Sydney usually a couple of times a week to speak to people who have been burnt, and to offer counsel and support. There's never a shortage of burns patients, he says.

For Karen, living in Queensland, as she does now, can be difficult because, where her skin has been badly burnt, the sweat glands have been destroyed. Overheating can be a problem. 'Until I found the Burn Foundation, I could not understand why I was hot all the time.' At her first retreat, she was complaining of the heat, and a fellow burns survivor explained the cause. 'My doctor thought I was going through menopause. She never said, "It could be your burns, Karen." I had been going to her for 25 years.'

Another health concern is that where the skin is scarred, it loses much of its capacity to stretch. Because Karen was a child when she was burnt, she has had to have a series of surgical releases so that her skin could accommodate her body as she grew. It's something she has to continue to be mindful of.

'Certainly, I couldn't afford to put on a lot of weight. I would be in trouble. My skin just wouldn't be able to cope with that now.'

Her skin is also more fragile. 'I get a lot of breaking of skin. I've got a particular one here,' she touches a spot on her breastbone with her fingertips, 'where it breaks open all the time and gets infected.' Such lesions can take months to heal.

Karen was five years old and home alone, playing with matches, when her dress caught fire. She saved her own life by running to the shower and turning the water on herself. She says that she's grateful her father wasn't home at the time. 'He would have put butter on my burns,' and, in Karen's frank assessment, she would have died. As it was, after quenching the flames, she ran out onto the street where two men found her and called an ambulance.

In 1968, as she remembers it, 'There wasn't a thing called rehab when I got burnt; you just went home and that was it.' Karen received extensive burns to her back, arms, neck, and torso. She lost a lot of her hair and still wears a wig.

Her parents sent her to some of the best schools in Perth, hoping, to little avail, that it would shelter Karen from the teasing and bullying of her peers on account of her scars. 'High school was a nightmare for me; it was horrible, really bad. I thought about taking my own life a few times, just because kids are so cruel.'

Karen never saw, let alone met, anyone with similar injuries as herself until 2002 when she saw on television Peter Hughes, a Western Australian man who had survived the Bali bombings. She remembers thinking, 'This is wonderful. There are other people out there like me.' Peter Hughes is now the patron of BFA.

When I ask Karen and Grazi what it is about being a burns survivor that is the hardest thing, they both answer without pause, 'The scarring.' Letting go of the fear of people seeing the scars is one of the freeing things that often happens at the BFA retreats, they tell me.

'It doesn't matter how you got burnt, or how old you were, or how it happened,' Karen says of the support offered by the BFA. 'You are a burn survivor, and we all there to help each other, and I just found that so inspirational. It just gave me so much confidence.' After having spent decades trying to hide her scars, she walked away from her first retreat in short sleeves, and with her hair tied back off her scarred neck. Twice, as we speak, Karen uses the expression, 'comfortable in my own skin'.

'The funny thing is,' Grazi says, 'when I see a new burn

survivor, I don't look at their face. You know what I look for? I want to see the scars.'

SCARLESS HEALING. It's a kind of mantra for Fiona Wood, the burns surgeon from Western Australia who barely needs an introduction. She repeats the phrase several times in the couple of hours that I am with her, and I've also read it in just about every article and profile of her.

To talk about burns is to invoke Fiona's name, carried as she was on the wave of media attention in the aftermath of the 2002 Bali bombings. Prior to that, she had, for years, more or less quietly gone about her business as a highly respected doctor and researcher in her role as head of the Royal Perth Hospital burns unit.

Born in the UK and married to an Australian doctor, she emerged during the Bali fallout as a kind of superhuman: she was a burns and plastic surgeon, she had invented 'spray-on' skin and co-founded a company to continue her research into burns treatment, and she was a mother to six children. She's been voted Australia's most trusted person more than once, as well as being both Western Australian of the Year and Australian of the Year.

Certainly Grazi Lisciotto is a fan, confessing that he is envious of the relatively little facial scarring carried by Peter Hughes, despite the time that Peter spent in Bali in between the explosion and getting medivaced to Perth for treatment. 'I wouldn't say he was lucky,' Grazi told me. 'You don't tell someone who's been burnt that they're lucky; but I wish I'd had that treatment.'

FIONA WOOD'S REPUTATION makes the prospect of interviewing her a little intimidating to me. I'm grateful when I turn up at the

Royal Perth Hospital that I don't have to present my own meagre resumé for comparison. She's not abrasive or overbearing but, in the face of her passion and her knowledge, I certainly want to avoid sounding like a fool.

She is shorter than I expected, her face broad and expressive, and her voice still has the trace of a Yorkshire accent. A crucifix at her throat is decorated with what might be an Aboriginal design in a mosaic effect. She is engaging, quick, and appears to have enormous self-belief, or at least she has an unwavering certainty that hard work and the determination to learn from whatever comes your way will bring its own rewards.

'This is the burns bible,' she tells me before I've settled in my seat in her office, pointing out a large red volume called *Total Burn Care* by D. N. Herndon. She gives me a list of other resources that I might find useful and I take notes and try to look intelligent.

She tells me about research that her colleague Mark Ferguson and his company Renova are doing into foetal skin and healing. Healed wounds in foetuses are virtually scarless, she tells me. Another fascinating thing regarding skin is that wounds in the elderly heal more slowly but are less likely to scar, she adds. Understanding the processes involved in the body's methods of regenerating and healing at different stages of growth and ageing is something that she and her team are currently very interested in.

'Puberty is a terrible time to be burn, for example,' she says. 'That hormonal time, when things are changing, and with growth spurts, we feel that's a negative time for the scar.'

'Our research,' she says, of her and her team, 'is very driven by clinical observation.' This is her strength; her many years of work at the coalface informs her methods and areas of investigation.

'I bring to the table experiences of how people have healed in different ways ... I've been in a position to manipulate that healing through surgical and non-surgical techniques local to the area and systemic to the whole body.'

Skin, she tells me, is 'the first line of defence, but it's not static; it's not a plastic bag that keeps our giblets in.'

The whole of Fiona Wood's professional life has been dedicated to understanding how best to encourage skin to regenerate rather than simply repair itself in response to injury. She believes that one of the keys to this is to engage the central nervous system in driving regeneration. 'We have a lot of evidence now that the quality of the scar is related to the quality of the neural function of the skin.'

Fiona's goal for any young doctor who comes her way is to turn them into a 'doctor squared'—a medical doctor with a PhD. 'Get 'em early and get that in their head.' This attitude extends to the nurses in her team, several of whom have done master's degrees. The combination of clinical medical skill and the capability to use that knowledge to further research is invaluable, she says.

Over the course of a couple of hours, she mentions an impressive number of areas that she and her team are researching, intend to research, would like to research, or are closely watching others research. These include the relationship between healing and ageing, personality type, and neural functioning; the question of whether suffering a major-trauma injury increases the risk of developing cancer; methods of lessening and improving existing scar tissue; pain management; the possibility of reversing cell death; and the various classifications of skin from all different parts of the body.

'There's so many things we could do,' she says. 'I'm getting more impatient as I get older.'

Wood, in conjunction with her colleague Marie Stoner, had already developed her spray-on autologous skin-culture technique (autologous meaning from the person's own body) when victims from the Bali bombing began arriving in Perth in 2002 with severe burns injuries. She and Stoner had been using the method for treating burns patients since 1994. Rather than culturing the skin cells taken in a biopsy for up to 21 days, and then applying it in sheets onto the wound, the skin cells are separated and put in a suspension that is then sprayed onto the injury. Back on the body, the cells begin to multiply, effectively using the body itself as an incubator.

This procedure can now be done within 30 minutes with a kit to produce a small number of cells that a surgeon can use within the operating theatre. A laboratory-based technique can also produce a large number of cells in around five days.

Not all hospitals in Australia have taken up the spray-on skin technique. It's not used at the Alfred Hospital, for example. Some burns surgeons remain sceptical as to whether it contributes to faster healing of the skin and less scarring.

'There's lots of ways of skinning a cat', is Fiona's diplomatic response when I ask why spray-on skin hasn't been universally accepted. 'But,' she goes on, 'every single time I've done a split-thickness skin graft [which consists of the epidermis and a small part of the underlying dermis]—and I've got some spectacular results—every single one of them has left a scar.'

'If we put the bar so low that we jump it every time, if my gold standard is split-thickness skin graft, why would I ever do anything different? If my gold standard is the normal skin of that body site of that person at that time, I can never achieve that with a split-thickness skin graft.'

I ask Fiona about the Victorian bushfires and their victims. 'Was your impulse to jump on plane and say, "Let me at them?"'

She drills me with a level gaze. 'You can't ask awkward questions.' And then, 'I would be very happy to help when I'm asked.'

IN FIONA'S VIEW, the area in which her spray-on-skin-cell technique has made the biggest difference is paediatric scalds. Before the technique was used, 92 per cent of paediatric burns remained unhealed after ten days, and required skin grafts followed by the wearing of pressure garments; now, her burns unit finds that only 20 per cent of these cases need that degree of care.

When they're presented with a patient, she and her team don't simply spray skin cells onto the site of a wound and consider their job done. 'It's a whole package,' she says. 'It's not just what's happening in the operating room; it's what happens in the dressing changes, it's what happens in the therapy as well.'

Techniques such as grafting and debridement still form part of the treatment regimen and, as in any hospital, her team's approach is multi-faceted: skin grafts, infection control, the rapid removal of dead tissue, exercise, reconstruction, pain management, nutrition, physiotherapy, scar minimisation, psychology, neural retraining, and the management of any other medical issues that the patient may have are all part of the equation.

When a burns patient presents, the first question that Fiona asks herself is: what is the regenerative capacity of this person, taking into consideration their general health, age, fitness, and nutrition levels? Then, what is the regenerative capacity of this wound?

Much of good wound care is commonsense, Fiona says—good

nutrition, good hygiene—but that doesn't mean that treatment can't be constantly improved. Her belief is that something can be learned from every patient to make her treatment of the next more effective.

Despite some of sections of the medical profession in Australia remaining unconvinced of the efficacy of spray-on skin, interest in the technology is growing elsewhere. After years of lobbying for a trial of her techniques to be done in America by the Food and Drug Administration, there is now a waiting list of US hospitals to be part of a trial, to which the US military is contributing funding.

I've accompanied Fiona to the University of Western Australia (UWA), where she is giving feedback to researchers who are refining a technique for interpreting the make-up of scars. On a screen in front of us is a 3D representation of a scar, produced by optical convergence tomography (OCT), which uses a wave of energy to image sections of the body.

Fiona holds a photograph of the same scar in her hand. 'You asked me before, why don't all people use skin cells,' she says to me, brandishing the photograph of a prominent welt left after the removal of mole high on a woman's chest area. In the photograph, the scar is mottled red and white, raised, with suture marks clearly visible.

Fiona is appalled. 'Why would you stitch it like that—with surface stitches?' The wound could have easily been stitched internally, she explains, leaving a much less prominent scar. Her face is a picture of outrage and incomprehension. She channels the careless surgeon: 'I've been doing it like this for 20 years, and I'll keep doing it like this for 20 years.'

The researchers at UWA are using OCT to assess scar tissue.

They hope that the information they gain will aid the minimisation of scarring.

To my untrained eye, the black-and-white images of the scar on the screen in front of us look similar to an ultrasound, and just as indecipherable. The researchers are decoding the differently shaded areas that give the scar its texture and shape: blood and lymph vessels, fatty lumps, fluid-filled blisters.

As part of their research, they need to manufacture a skin-substitute to use as a control to test the accuracy of the images that they're producing. In a half-circle around Fiona, they discuss the various substances that they have used, and present to her scraps of fake skin for her to comment on. She's across all the areas of their research: the chemistry, the professional networks, the techniques and technology, the theory, the practice, what's been published, and by whom. Her audience drink it all in: her knowledge, her passion, her vitality.

Prompted by the fact that albumen has been used to make the skin substitute, Fiona talks about a researcher in the US whose work she knows. He is working on reversing the chemical processes that occur when poaching an egg; it's all about reconfiguring proteins. If you can un-poach an egg, then perhaps you can un-burn burnt skin.

'Push better' is another of Fiona's mantras. 'We can always learn, always use the information we gain to do better next time.'

I'm left in no doubt that those who come into her orbit without that attitude are given short shrift. She wants those whom she works with to step outside their comfort zone, and to constantly engage in constructive questioning and problem solving. 'I cannot sit back and think, "Well, this is as good as it gets."'

Since Bali, the number of burns patients admitted to Western

Australian hospitals has dropped significantly, along with the average of total body-surface area that burns patients present with. Somewhat surprisingly, this is not because of a whole scale burns-prevention program, but rather because of the burns service's efforts in being (as Fiona puts it) 'rabidly aggressive in teaching everybody who'll stand still long enough' the basics of burns first aid. Fiona believes this has done much to reduce the severity of burns in the general population, because this knowledge has become embedded in the community as well as raising awareness of the risks of burns.

And how does Fiona feel in those rare times when she pauses at the edge of the abyss that is a severe burns injury? 'I have a really clear understanding that your life can change in an instant,' she says, 'because that's what I see on a daily basis.'

Whether that accounts for her relentless drive and energy to understand the skin, its functions, and how to manipulate its regeneration is moot. All she knows is, 'life is an extraordinary gift that I think a lot of people waste.' She is determined that she won't squander the years she's been granted on the planet, not personally and not professionally.

One of the ways that she deals with the inevitability of patients who have 'just slipped through our fingers despite everything we've done' is by being determined that every experience inform her methods of treating the next person with a burns injury: 'I have people that I will never forget, as I've said, that have died, but I know they have influenced the lives of all those people coming subsequent to them.'

'The way I govern day to day is to make sure we learn something every day. And I try to teach that to the youngsters. Even if you walk away from an operation, how could you do it better tomorrow? Not because you've done it badly, but because we should always be learning, we should always be refining, always be pushing just that little bit further so that the quality of life we give that person is the very, very best.'

I wonder how Fiona, and health professionals like her, bear it—dealing with the lives that slip through their fingers. Echoing Yvonne Singer, Fiona says of her patients: 'The ones that keep me going, if you like, are the ones who are just extraordinary in how they can take on that suffering and push back and grow. They really are a lesson for us all in that, if that happened to me, I hope I can behave that well.'

I HAVE GONE as close to the chasm of a severe burns injury as I want to get. I've seen some of the courage and compassion that it takes to stand on the edge, and the strength required to begin the climb back out. Yet, having spoken to those who have survived the fall—and those who tenderly and doggedly coax, haul, and care for them as they pull themselves up—my dread of the possibility of that fate befalling me has receded infinitesimally.

Leaves on the Tree of Life: the Donor Tissue Bank of Victoria

The only gift is a portion of thyself.
— Ralph Waldo Emerson, 'Gifts', 1844

ONE THING THAT I learned from speaking with Yvonne Singer at the Alfred Hospital and Fiona Wood is that the best dressing for burns is more skin. It seems obvious when you think about it; after all, hospital burns units routinely graft a patient's own skin to help with the healing process. Would the skin of another person be able to fulfil a similar function?

In the case of third-degree burns, the skin won't repair without a graft because not enough tissue remains for the skin to heal itself. However, when someone has suffered a burn to a large proportion of their total body-surface area, an autologous skin graft is not always possible. They simply may not have a large enough area of undamaged tissue remaining from which a graft can be taken.

Skin, however, is an organ, and it can be donated in much the same way that other organs, such as kidneys, lungs, and hearts, can be made available to living recipients from recently deceased donors.

There are some differences, of course. Skin is one of the most antigenic tissues, meaning that it will usually provoke a strong immune response, and the recipient's body will eventually reject the epidermis part of the graft, at least. Another difference is that for other organs to be donated successfully, the donor, while brain dead, must be on life support to ensure that their organs remain viable. Skin, however, can be harvested successfully until 48 hours after death, even after the donor's other organs have ceased to function.

An allograft (tissue from another individual) can be used to treat any area that has suffered a loss of skin: burn, trauma, chronic ulcers, epidemia bulosa. In Australia, however, the vast majority of donated skin is used in the treatment of burns.

ON A BRISK MID-WINTER afternoon, I visit the Donor Tissue Bank of Victoria (DTBV), the only multi-tissue and skin bank in the country. It exists within the Victorian Institute of Forensic Medicine (VIFM)—the same laudable establishment through which I contacted pathologist Shelley Robertson—which, in turn, makes up part of the larger complex that is the Coronial Services Centre of Victoria (CSC).

The DTBV is the only tissue bank that exists within a coronial system in Australia, and it has a unique relationship with both the VIFM and the coroner. The autopsies performed at the CSC not only gather information relating to causes of death, but also provide an opportunity to access human tissue for donation purposes. It's from here that hospitals like the Alfred can source allografts for patients whose burns injuries have left an insufficient amount of their own skin suitable for grafting.

The DTBV site is made up of what is called 'the core'—the

facility where retrieved tissue is processed and handled, microbiology and serology labs, a mortuary, an autopsy room, offices, and storage areas. It is a highly regulated environment with all of its operations dictated by a multitude of procedures, regulations, and processes of validation. It must be so for an institution that deals with the sensitive task of asking the grief-stricken to consider donating the tissue and organs of their loved ones, and the technical issues of retrieving, processing, and storing that tissue in a respectful and non-hazardous manner. When the death that has brought about the opportunity to harvest tissues has been caused by violence or another unexpected means then the sensitivity and care required is even greater.

THE FOYER of the DTBV is a nondescript space with the prosaic, utilitarian feel of a government facility. A receptionist's desk and the muted furnishings are unremarkable, although I do register the presence of a large painting on the wall as I sit in the chair beneath it waiting to be ushered through to the offices of the Tissue Bank. My focus is on what I hope to discover beyond the doors that lead to the inner rooms of the DBTV.

I'm shown through to the office of Kellie Hamilton, senior scientist at the DTBV. She is glossy with her burgeoning pregnancy, brimming over with health and good humour, and for a moment I'm slightly taken aback by such an abundant display of wellbeing within the walls of an institution largely concerned with cadavers. Hamilton has worked at the Tissue Bank for 15 years, beginning in retrievals and processing before moving to research and development.

Marisa Herson, head of the DTBV, is also present. Older and smaller, she is dressed all in black, in a sober shirt and trousers.

Both have set time aside to talk me through their roles and the functions of the institution that they work for.

Marisa is a trained burns and plastic surgeon who is experienced in reconstructive surgery and has been head of the Tissue Bank for two-and-a-half years. She was born in Brazil and, among her previous roles, was the director of a hospital burns unit there.

Both women are welcoming and urge me to ask any questions I like, yet I sense a wariness in Marisa, and I guess she may be somewhat suspect of my motives. I scribble copious notes to keep up with the technical information that they give me in response to my questions, mildly put out that Marisa gave a polite but firm 'no' to my request to record the conversation. She is helpful and smiling, but I pick up her fierce protectiveness of the Tissue Bank, its reputation, and her responsibilities to those whose trust she is obliged to honour.

The skin bank has occupied a vivid place in my imagination since I first heard about it. I envisaged it, fancifully, as a vast glasshouse where acres of skin billow like parachute silk, weightless and transparent under its translucent roof. It would have lights that simulate warm spring sunshine, and a fine spray of water that keeps the skin pliable and moist. Technicians in white jackets and face masks would tend the tissue's growth like sanitised farmers in some bizarre distortion of the rural idyll.

This is, of course, an entirely whimsical vision, but it has entertained me, and I have resisted discarding it altogether. I am hoping to see the real thing, to be able to finally see skin in the way that I have been thinking about it: as something in and of itself. I am to be disappointed, however—Marisa makes it clear almost immediately that this won't be happening today, and I must be content with asking my questions and listening to the answers.

Skin can be harvested from just about anyone—once they are dead. (I recall Yvonne Singer's dislike of the term 'harvest'; but, for me, it has a wholesomeness that alludes to the hoped-for positive outcome of the process.) Although skin can be successfully retrieved from patients who are already deceased, there is one factor that can rule out its suitability for donation: the donor's age, 65 to 70 being the upper limit.

Skin removal may sound like quite a drastic procedure, but Marisa and Kellie both agree that it is the least invasive of all organ-donor procedures. Contrary to a widely held erroneous assumption, says Marisa, skin donation does not result in the entire body being flensed. The layer of skin that is taken is very thin so that, once the procedure has been completed, the site wound will look similar to that left by a slight graze.

On a subsequent visit, Marisa shows me a cartoon by Claude Serre, who is known for his sometimes graphic and macabre images. It shows a flayed man, his muscles and sinews exposed, being presented by a shop assistant with a skin suit on a hanger for his consideration. The pelt is displayed in the same manner that a gown might be shown as a shimmering confection of silk to a woman contemplating its purchase. Behind the two figures is an unmistakeably female hide hanging on a rack. The cartoon amuses Marisa, but it also highlights for her the misapprehension that she fears is prevalent in the wider community, and that contributes to a less-than-positive perception of skin donation.

The procedure for the retrieval of donor skin is similar to that performed in hospitals to harvest tissue for an autologous skin graft. It's what precedes the retrieval that marks the process as essentially different, and that is the altruistic act of a family member agreeing to the removal of skin from the deceased's body so that it may be used to benefit another.

It is not a small thing to ask in any circumstances. Given that most of the skin sourced by the Tissue Bank is through the coronial process—that is, from people whose deaths are being investigated because of the circumstances surrounding them (often sudden, traumatic, or unexplained)—the gravity of the task is even greater. It is a responsibility that those who work at the DTBV bear mindfully. Largely out of respect for that initial act of altruism, there is strong cultural resistance within the VIFM to any move to commercialise organ and tissue donation.

The first step in the process of retrieval, after consent has been given, is a physical assessment of the skin. The overriding consideration when determining whether a donation will be sought and accepted, Marisa tells me, is that the 'benefit should surpass the risk'; the risk being the possibility of transferring infection and disease from the donor to the recipient. The skin must also be largely intact and, given that a significant proportion of the deceased subject to the Coroner's interest may have suffered trauma or serious injury, this is not always the case. The other non-negotiable factor in the retrieval process is that it must be respectful of not only the deceased's family but also of the deceased themselves.

Following disinfection to remove any flora, the skin is retrieved in a secluded area of the Tissue Bank using a device called a dermatome, a surgical instrument with a very fine blade that removes the skin in thin strips. The technician removing the skin is scrubbed up and gowned as in a conventional surgical procedure—and in a photograph, I can see that it does look similar to a normal hospital surgical suite: the gowned technician, the scrubbed and shining surfaces, and the operating table on which the deceased is placed.

The air in the core is controlled to keep it as pure as possible to prevent infection and contamination. Split-thickness allografts are removed from the body. The strips of skin are usually taken from the back, the lower limbs, and the back of the arm, sites dictated both by technical considerations and sensitivity to the deceased's family: the strips of skin obtained must be large enough to be used as grafts and taken from areas that won't be visible if the deceased is viewed by their relatives, or is to be dressed in a particular manner for burial. Hands and faces, for example, would not be used as sites from which to take skin. Often, too, relatives may ask that particular areas be left untouched, such as those marked with tattoos. At all times, both Kellie and Marisa stress repeatedly, the wishes of the family are followed.

I begin to ask a question about the ramifications of grafting skin of a different hue onto a body, and then stop when I realise the question is foolish. Marisa has already mentioned that the epidermal section of the graft will slough off some weeks after grafting, as the recipient's body will eventually reject the foreign tissue. Marisa has anticipated the direction of my abandoned question, and tells me that dermis is a similar colour in all races: the melancytes, which produce melanin and so determine the colour of the skin, are present in the epidermis only. The large amount of collagen in the dermis makes it less antigenic, and it is possible for the recipient's body to integrate it with its own tissue.

Unlike other organ donation, the skin does not need to be cross-matched with the person who is receiving it because, essentially, the use is temporary. Its role is primarily as a dressing to prevent bacteria getting in and electrolytes seeping out. Marisa compares a burn to a house losing its roof: by putting on a temporary roof in the form of a piece of harvested skin, the 'housewife', she says, has time to clean the house and prepare it against the storm of possible infection.

Harvested skin can be kept viable for up to five years when stored frozen in the vapour from liquid nitrogen at between -130 and -196 degrees Celsius. The retrieved skin is placed is an aluminium bag that can withstand the extreme conditions, and then suspended in the vapour within large Dewar flasks (double-walled flasks of metal or silvered glass with a vacuum between the walls, used to hold liquids at well below ambient temperature), which are linked to an alarm system that picks up any deviations of temperature. Anyone who has defrosted a frozen tomato will have witnessed the damage that can be done to a cell wall in that process. As the tomato freezes, the liquid within the cells forms crystals. The tomato turns to mush when it defrosts because the spikiness of the crystals ruptures the cell wall, allowing the liquid to leak out, and irreparably damaging the tissue.

At the Tissue Bank, through a cooling technique which utilises a substance that reduces the spikiness of the crystals as the cell-fluid freezes, the tissue can be defrosted with minimal damage.

Towards the end of my visit, Marisa has to leave, and I finish off my interview talking to Kellie. I mention that I'd like to see within the core of the facility and, if possible, where the skin is stored. Kellie is confident that this can be arranged, but Marisa has to be consulted, so I leave with a promise that she'll contact me and let me know a time when I can return. A few days later, I learn that Marisa is reluctant to let me see inside the facility; but, with the help of Deb Withers, the public-relations consultant, Marisa offers to give me a presentation of a more in-depth view of the Tissue Bank's activities.

On my second visit to the DBTV, Marisa meets me in the foyer. She seems less wary this time, more relaxed, although I am disappointed to learn that my 'tour' must be a virtual one. I had imagined being able to see the actual harvested tissue, but Maria tells me that once the Dewar flasks are opened the skin would have to be destroyed.

'We are stewards of a gift,' Marisa says, in explaining why she can't allow me to see any retrieved skin. There is an expectation held by the donor's family, she explains, that the gift of tissue will be used for transplant, and to let even part of that donation be lost through allowing me to see it would be to dishonour that gift. Put this way, I can't but agree with her. In addition to her responsibility to the families of the donors, there is also the imperative of keeping the facility as sterile as possible. This is not only for my own safety and that of the people who work here — the DBTV does work with potentially bio-hazardous material, after all — but also to ensure that the donated tissue is not contaminated by handling, or unnecessary movement through the facility.

As Marisa takes me on the virtual tour of the Tissue Bank and the work that they do, she occasionally apologises for the images on the computer, warning me in advance and checking that I am comfortable with the slides that come up on her computer screen. There are pictures of raw wounds — some open and gaping — organs, and bodies that show the ravages of injuries and the effects of surgeries that have repaired them, or attempted to, sometimes imperfectly.

'Close your eyes, close your eyes,' she warns as particularly harsh images appear, using yellow Post-it notes to cover up those that she deems too strong for my uninitiated eyes.

Some of the images are disturbing: strips of grafts over a large wound look like large bandaids; a girl who was scalped in an industrial accident, losing an ear as well as her hair and the skin on her head where it grew. Many of the images come from Marisa's time in Brazil where she headed up a burns unit.

A photograph of a strip of retrieved skin is on her screen. It is, she tells me, exactly what I would see if she took me into the facility to see the real thing. It looks unremarkable, like a strip of wet, grey paper. The thickness of the retrieved skin is similar to that taken from the skin for autologous grafting. The epidermis and a small part of the dermis is harvested during the retrieval, and has a depth of approximately 0.40 cm. Depending on the part of the body that the skin has been retrieved from, the skin could be up to 20 cm long, and is usually between four and 10 cm wide.

As the time for my visit comes to an end, Marisa tells me about Leaf Day, which the Tissue Bank holds every year. There is a large mural of a tree in the foyer, she tells me, have I seen it? I have to confess that I have not even registered its presence. She suggests I look at it on my way out. At the end of each month, Marisa explains, leaves bearing the first name of each tissue donor are placed on the mural. After 12 months, the tree is covered in leaves. Towards the end of every year, usually in November, all the families of the people whose names appear on the tree are invited to attend a function at the Tissue Bank with tissue recipients and the researchers, technicians, and scientists who work there. Over morning or afternoon tea they tell their stories.

Marisa urges me to come to the next one to be held in a couple of months, and my ungracious and clumsy response is that my deadline will have passed by then. 'Not for your book,' she reproaches me gently, 'but for yourself. It is an immensely moving day. Difficult and beautiful.'

Marisa walks me out to the foyer, and I notice that she touches me as we say goodbye; just a slight extension of her hand to brush my arm. I watch her leave, and then look for the mural. There it is: a stylised tree on a hillside with spirals of paper leaves in green, red, and yellow that seem almost to spin like pinwheels on the surface of the painted image. Beneath the painting is a brass plaque with the words: *In memory of Ben. Donated by Mears Family and Friends.*

I had sat beneath the mural twice, and not even looked at it. Now, I take the time to read the names of those who have died and donated their tissue for the benefit of others. Like the occasion of Leaf Day, where the families of the donors, the staff, and the recipients come together to tell their stories, to consider these lost lives and the ones they have saved, contemplating this simple tribute is difficult and beautiful.

Second Skins: semi-living sculpture

Garments are humans' fabrication and can be explored as a tangible example of humans' treatment of the Other.
— Tissue Culture and Art Project

THE SKIN IS a potent subject for artists. As a metaphor, its meanings and allusions are myriad and shifting, its resonances rippling through questions of identity and gender. Skin is a permeable and changeable membrane, a defining boundary that delineates us from our surroundings, and a site of communication. As such, it is an ideal vehicle through which to engage with and examine questions of beauty, the self, revelation and concealment, materiality, culture, and technology.

Given its potential to regenerate, its pliability, its tactility, its tendency to form into folds and creases, skin is a favourite subject for artists to explore through the traditional mediums of paint and ink, particularly given the striking analogy between the surface of the canvas and the surface of the body. It has enjoyed no less attention in the fields of sculpture, photography, and textiles.

The nude—the portrayal of the body clothed only in its own skin—has been a popular subject with artists for centuries, and with little wonder. Few topics provide the frisson, the eroticism, and the sheer voyeurism of the nude.

Related to this exposure of the body are the ways in which we drape and conceal it. It has been millennia since the only function garments served was to protect our fragile hides from the sometimes-pitiless nature of our external environment. Clothes can alter our shape, draw attention to or away from the various parts that we wish to emphasise or disguise. Fashion is now both defended as an artistic pursuit, with its high priests and priestesses feted as creative geniuses, or derided as facile and misogynistic nonsense.

In the last decade or so, with tissue engineering now a reality, and the technology available to artists who seek it, skin and other human tissue may be cultured and manipulated and can, itself, be the medium. Our flesh, and the materials we use to sheath it, can become entwined—literally.

THE JACKET is fleshy-pink in colour and itty-bitty-doll-sized. It hangs, suspended by thin filaments, within a rubber-stoppered glass flask that is just one part of the system that keeps the tiny garment alive. Yes, alive—or rather, in the lexicon of its creators, the artists Oron Catts and Ionat Zurr, semi-living, for the material out of which the jacket is fashioned is not cotton, wool, or linen. Neither is it a blend of viscose, rayon, or polyester. It has, in fact, been grown out of immortalised cell lines, which form a living skin of tissue over a biodegradable polymer matrix that has been shaped into the form of a miniature jacket.

Immortalised cell lines are cells that, through mutation or intentional modification, are able to replicate themselves indefinitely. They are used extensively in medical research and biotechnology.

The jacket, in effect, is fashioned from a form of leather. More specifically, Oron and Ionat have dubbed the work, in a term that is heavily weighted with the ambiguity and irony of the jacket's precarious existence, 'Victimless Leather'. After all, this scrap of ever-dividing mouse and human cells can only exist on its polymer base within a highly specialised environment, fed with a serum containing foetal-calf blood. Removing it from its uncontaminated habitat, or simply touching it with an unsheathed finger, would prove lethal to it through the introduction of bacteria.

The flask in which the jacket lives is connected to other glass vials and a small pump by tubes that funnel nutrients to the jacket, and moderate the air within the flask to keep the conditions conducive to its continued existence.

Oron compares the system supporting this not-quite-living oddity to a body: the gas-exchange chamber acts as the lungs, the pump is the heart, the vial containing the pinkish transparent serum is the stomach, and the jacket itself is an organ. He tells me that, since they first created the piece, it has 'taken on a life of its own', becoming an iconic work whose meaning in the wider world they, its creators, have little control over.

Ostensibly, 'Victimless Leather' flirts with the future possibility of being able to grow a wearable leather jacket using cells cultured from our own bodies, or perhaps from immortalised cell lines. After all, we are the naked ape, and must clothe our own poor, thin hides in something. This way, no animal would need to die in order for us to drape our skin in a second skin.

It would be a mistake, however, to assume that this is where the meaning of the work begins and ends. Like the skin itself, 'Victimless Leather' is a multi-layered construction with a range of functions and meanings. Through this deliberately provocative creation, the artists are seeking to unsettle, and to gently, yet firmly, encourage their audience to articulate, at least for themselves, the ethical dilemmas and sense of unease that it generates.

This kind of provocation is central to Oron and Ionat's greater work, the Tissue Culture and Art Project, which involves ongoing research into the use of tissue technologies for artistic expression. The artistic terrain they explore is rich in narratives that strive to interrogate the use of biotechnology and the exploitation of life by humans; a place where tiny flesh-coloured worry dolls, grown from polymer and living cells, hang suspended in glass tubes, inviting onlookers to confide their anxieties, and where pigs might grow wings.

Oron and Ionat's interest in living tissue extends beyond skin: their art is concerned with life and the nature of the 'semi-living'. To engage with their work requires a re-adjustment; a willingness to move beyond the 'yuck' factor and the temptation to trivialise it as mere whimsy. They are serious and thoughtful artists who regularly confront their own disquiet, but who, nonetheless, enjoy and explore a playful perversity.

They were the first artists-in-residence to be appointed to the Harvard Medical School, where they were given the same rights and responsibilities as other research fellows. Oron is also the director of SymbioticA, an art and science collaborative research laboratory at the University of Western Australia (UWA), and Ionat is the laboratory's academic coordinator. In 2010, Oron was also appointed visiting professor at the Royal College of Art in London.

I met with Oron and Ionat during the same trip to Perth in which I met with Fiona Wood, and I was delighted to find out that they know each other. When Oron was completing his honours thesis in the mid-1990s, in which he was attempting to grow a skin of living cells over inanimate objects, Fiona, with her extensive knowledge of cell culturing, was one of the people whom he interviewed. (Fiona told me she still has Oron's thesis somewhere in her office. She recalls walking out of the room where they first met thinking 'I'm so conservative.' Not, I imagine, a thought she is familiar with.)

As artists who engage fully in the laboratory processes that are required of their work, and for whom a basic tenet is 'tissue is not an abstract material', both Oron and Ionat appear to feel a sort of kinship with Fiona. The creative and exploratory nature of Fiona's research is a point of connection for them, and even Oron, with his restless urge to provoke, seems slightly envious of her energy. When I mention the sections of the medical profession who remain sceptical about aspects of Fiona's research, Oron cites it as evidence of the inherent conservatism of the scientific professions.

'Doctors are not considered to be scientists,' he says, 'so when they come up with something new, the scientific world rejects them.'

Ionat agrees. 'The surgeon is a bit like an artist,' she says. They both begin from a point of trying to provide a solution to a problem, as she sees it. Scientists, she feels, tend to begin from a different starting point altogether. Yet, here, Oron and Ionat find themselves in a laboratory cheek-by-jowl with scientists, for SymbioticA exists within the Department of Anatomy and Human Biology at UWA.

SymbioticA manages, to some extent, to fly under the radar in what many regard as one of the more conservative universities in one of the most conservative states in Australia.

Dealing with the shifting perceptions of life brought about partially through the advances in life sciences and biotechnology, SymbioticA provides artists with the opportunity to use the tools and the technologies of science in their art. It thereby gives them the skills to go beyond merely commenting on science and biotechnology to actively engaging with the actual processes and materials that these areas of knowledge rely on for their advancement. Artists completing residencies at SymbioticA, and students wishing to work in this area, can learn the skills required to culture cells, infect and manipulate tissue, and safely handle viruses.

SymbioticA's laboratory looks like a regular scientific place of experimentation: the benches are crowded with the paraphernalia of discovery, including microscopes, glass vials, rubber tubing, clamps, pipettes, and forceps.

Oron shows me through the lab with diffident care, pointing out the small jacket-shaped mass at the centre of a conglomeration of glass and tubing that is 'Victimless Leather'. It is soon to travel to Luxembourg for yet another exhibition. The hardware will travel, that is; the little jacket at the heart of the exhibit will be regrown once it arrives.

Travelling with biological material is fraught with technical, and sometimes legal, difficulties, and they rarely do it. As is often the case, once the exhibition is over, their display will be gently executed, perhaps by inviting the audience to touch the collection of dividing cells.

'Victimless Leather', which Oron and Ionat created before SymbioticA was established, has been shown in a range of different contexts: as science, as design, and as political art.

'We never thought that it would be so big, or generate so much attention and discussion,' Oron says of the work. Articles about it have even appeared in fashion magazines. 'Which,' he says, 'we find really disturbing.'

In the early shows, he says, they were deliberately using the rhetoric of 'victimless'. The fact that the serum that provided the nutrients for the cells came from animals was not mentioned, for example.

He recalls a question-and-answer session following the exhibition of the work at a conference when a participant, outraged by their lack of heart, as he saw it, challenged them on their ethics. He happened to be dressed completely in leather. 'You're wearing cow on yourself,' Oron replied, 'and you're asking us where is the heart in our work?' It was almost too perfect, he says, with some people in the audience assuming that the leather-clad man and his question were part of a deliberately staged set-up.

Oron and Ionat relish such exchanges. Their work is intentionally manipulated to evoke reaction; even still, Oron appears bemused by the fact that people seem more comfortable with slaughtering animals to exploit their skin for clothing than they are with thought of living cells being cultivated in the shape of a jacket.

People are so blind to traditional modes of exploitation, he says, that they have to be reminded where the leather in their shoes comes from.

However, Oron and Ionat's intention is not to tell people what to think, or even to reveal their own position on the questions that their art provokes.

'As artists,' Oron says, 'we are interested in the issues we are raising rather than with our opinions. We deliberately engage in a very ambiguous presentation of our work, which is why we employ irony—which in most cases people don't even pick up. We're trying to create this ambiguous zone where people are being challenged and need to make up their own minds rather than responding with what we tell them to think.'

This approach often invites anger, but Oron is unconcerned. 'We are the messengers,' he shrugs. 'You shoot the messenger.'

Far more radical assaults on notions of life and its sanctity are occurring everyday in the laboratories of agribusiness and the biomedical and pharmaceutical industries, he asserts. Large-scale industries hide their methods of research and production, which routinely manipulate living tissue behind utilitarian arguments that the ends—feeding the hungry, healing the sick—justify the means.

Oron dismisses such justifications as being largely 'bullshit'. Millions of people die of preventable diseases and starvation every year, and it is mainly due to a lack of political will and the desire for profit, he says. 'Genetically modified food is not going to feed the world. In affluent countries, farmers are destroying food because they do not want to sell it cheap.'

Art is perceived as being frivolous, he says, and artists can't hide behind utilitarian arguments about its purpose. 'We expose the ethical questions in their fullness because of the perception of this frivolous activity. And that's exactly why we're interested in doing this.'

Viewing their work can be uncomfortable. The contemplation of these semi-living sculptures—the tiny, mouse-sized jacket, a possible second-skin of skin; the fleshy little worry dolls with their blank faces, spongy and potent; and wing-shaped objects grown out of living pig tissue—makes my own skin twitch in the way of a horse trying to shrug off a biting insect. Is it a concern that life is being manipulated for art? Is it an abomination to combine the cells of mice and humans in a growing, living object? Does it matter if living cells, and the life within, become just another product for exploitation?

In keeping with their appreciation for ambiguity and irony, Oron and Ionat are chuffed when their work doesn't always go as planned. In 2008, 'Victimless Leather' was exhibited as part of the Museum of Modern Art's show 'Design and the Elastic Mind' in New York. The cells forming the skin of the jacket began migrating to other areas of the system, multiplying at such a rate that the incubator became clogged. In the end, the curator had to shut down the nutrient supply to the jacket, prompting the headline in the *New York Times* 'Museum Kills Live Exhibit'. It still makes Oron smile gleefully as he recounts the story.

Oron and Ionat often include the death of their creation in their sculptures, in what Ionat calls 'evocative killing rituals'. Audience members are asked to touch the conglomeration of living cells, inducing 'a gentle type of death'.

It's another layer of irony, says Ionat. 'When you think about skin, it's a lot about tactility and touch, and with those things [their sculptures]—a touch means death, in a way. So it's always mediated by technology.' Not to mention issues of health and safety, as Oron points out.

'So it's always behind barriers,' Ionat continues. It is not until the moment of their tender execution that their sculptures of tissue can be touched.

Oron began as a designer before deciding that it was as an artist that he would be able to successfully explore the ideas he was interested in. One of his and Ionat's early undertakings, which began as a design project as a basis for his thesis, was growing a skin of living cells over glass objects. They were interested in the concept developed by the Italian design theorist Ezio Manzini that objects are entities in themselves, similar to plants in a garden, that require our care, and they wanted to investigate how people's relationships with inanimate objects might change if they were semi-living. The knife-edge that the project rested on was whether having a living skin would make people more caring towards the objects, or would simply contribute to their objectification of life as another product ripe for our exploitation.

In order to learn the laboratory techniques required for their project, they began working with a researcher at the Lions Eye Institute in Perth who was developing an artificial cornea. The skin that grows over the eye has no dermis but is made up of a very thin layer of epidermal cells. The researcher was developing and testing materials suitable for creating an artificial cornea, and then attempting to grow epidermal cells over these materials.

As Ionat and Oron tell it, their first encounter with the researcher whom they were going to work with was promising. When they walked into the lab, he greeted them with a vial of rabbits' eyes in his hand, and a quote from the futuristic movie *Blade Runner*. They got the reference immediately. It comes from a scene where two replicants — engineered humans with a pre-programmed mortality date — come looking for answers in their quest to put a stop to their death clock. The technician in the movie, fearful of their quiet menace, and in answer to their demand for information about their morphology, their programmed longevity, and their 'incept dates', tells them, 'I don't know such stuff. I just do eyes, just eyes.'

For the two artists, it was a good omen. 'We were so happy,' recalls Ionat.

They soon found themselves tackling the visceral and confronting tasks involved in using living cells. 'We spent a lot of time ripping out skin from rabbits' eyes,' Oron says, and then, after a pause, 'after they'd died, obviously.'

Animals that had been killed for consumption in restaurants were the source of the eyes that they used in the lab. The rabbits would be killed in the morning, and the heads sent to another research facility, where the skull would be cut open and the brains removed. The artists would receive the bisected rabbit heads from the first lab, still with the eyes intact, that same afternoon. After removing the eyes and placing them in a solution, they would be kept overnight in a fridge. The next morning, more than 24 hours after the animals had died, the artists would remove the skin from the eyes, which would still contain viable cells.

Ionat and Oron then designed a series of glass figurines over which they wanted to grow the skin cells. They took the living tissue from the rabbits' cornea skin, and placed it into a hole on the surface of the glass object. The skin cells began to migrate out of the hole, and eventually completely covered the figurines.

During this time, they also examined the social and philosophical implications of their project. They were interested in 'the gap created by the slow and non-purposive biological evolution as opposed to the fast and goal oriented technological one'. They concluded that 'our cultural and social values are not yet equipped to deal with the problems that this technology presents to us'.

Another aspect of Manzini's ideas that they were interested in was that semi-living objects, because of the care that they need to survive, would result in less waste. The theory was that investment in an object that was also a living thing would translate to a less throwaway culture, one in which objects were kept and treasured. Instead, they were soon confronted with the reality that tissue culturing is an extremely energy- and resource-intensive operation.

Ionat makes a small noise of incredulity and dismay: 'The amount of waste we create …'

SymbioticA offers residencies to other artists who are interested in exploring and developing ideas around tissue culture. Oron and Ionat are at pains to point out that these works are not collaborations: the residents are independent artists whose vision they help to facilitate. They do this through teaching them the skills and assisting them to access the technologies that they require to realise their artistic project. Several of these resident artists have been interested in exploring and creating artworks that incorporate their own or other people's skin cells.

In 2007, ORLAN, an internationally renowned French artist who has worked across video, photography, sculpture, and performance, completed a residency at SymbioticA. ORLAN is perhaps most famous for her work in the early 1990s, when her own body became the material for her art and she conducted a series of surgery-performances, altering her appearance through plastic surgery.

'She is the most French person one can meet,' Oron says. 'To a large extent, even her surgical performances were based on *laïcité*.' (*Laïcité* is the French concept of multiculturalism and secularism. One manifestation of its implementation through policy has been the prohibition in France of the wearing of religious insignia in schools, including headscarves and crucifixes.)

ORLAN has a longstanding interest in hybridisation. Her aim, initially, for her residency at SymbioticA was to transplant skin from the people of other races onto her own skin to create a harlequin effect. Oron and Ionat had to break it to her that for her vision to be realised—diamond-shaped patches of different coloured skin covering her flesh—she would need to be on anti-rejection drugs for the rest of her life.

Oron and Ionat also had ethical concerns about the project as, essentially, it would require wounding other people's bodies in order to make the patchwork skin—biopsies would have to be done in order to harvest it. Also, Ionat says, ORLAN, like many artists who come to SymbioticA, had somewhat naively accepted the hype of the biomedical world. Her ambitions of culturing full-thickness skin from other individual's and grafting it onto her own body were unrealistic.

ORLAN's concept of race, and the idea of race in general, was also problematic for them. Modern biological science has brought the way race is thought of into question. DNA technology has shown that the genetic differences in individuals within so-called racial groups can be greater than those between individuals from different racial groups.

It took several years of discussion for SymbioticA and ORLAN to come to an agreement about what was possible, and what would comply with the requirements of UWA's ethics committee. ORLAN modified her original idea of grafting other people's skin onto her own body to the no-less-flamboyant project of creating a harlequin coat 'made from an assemblage of pieces of skin of different colours, ages, and origins'.

'She insisted that she would get a biopsy,' Oron says, 'and that was reasonable. We have quite a few ethics clearances for artists who want to use their own tissue to go through the self-harm process of a biopsy. As long as they're consenting and aware, that's no problem.'

There was still the issue of finding a source for the other skin cells that ORLAN wanted to culture along with her own. Oron and Ionat quickly quashed ORLAN's initial idea that, as she was creating the project within Australia, she could culture the skin of an Indigenous Australian to grow along with her own. Their facial expressions and exchange of glances reflect their remembered horror of the possible fallout of ORLAN's request if they'd failed to point out the political and cultural minefield she would have stepped into had she tried to proceed. The solution was to obtain tissue from the American Tissue Culture Collection (ATCC), a global organisation that provides human and other tissue for research purposes. And it's available through mail order.

It was agreed that it was acceptable for ORLAN to obtain tissue from the ATCC for her project. Obtaining cells from this source meant that no one, apart from herself, would need to be injured to obtain tissue.

Oron concedes that there may be issues of consent, as the people who have agreed to have their tissues stored and used for research may not have wanted, or consented to, them being used for an artwork. The tissue ORLAN chose originally came from an African-American female foetus at 12 weeks' gestation. The ATCC still use ethnic characteristics to categorise some of the tissue that they store.

The photographs on the SymbioticA website of ORLAN preparing for the biopsy of her skin to be taken have a self-conscious theatricality to them that borders on the comic. Her theatre gown is a voluminous sheath patterned with coloured diamond shapes, as is the sheet and pillow on which she reclines, waiting for the surgeon's scalpel. In her left hand, she holds a book by Michel Serres, a French contemporary theorist who uses the harlequin as a metaphor for multiculturalism. Her shock of hair—black on one side, white on the other—evokes Cruella de Vil, the evil dog-napper in *One Hundred and One Dalmatians*, and the surgical implants on her forehead gleam whitely above her large eye-glasses—bumble-bee black and yellow.

After the biopsy, Oron tells me, an event witnessed by photographers and other interested observers, ORLAN went dancing, tearing open the wound left by the biopsy low on her hip. 'She's an amazing person, ORLAN,' he adds, somewhat redundantly.

But what was the final work like? I ask. I have imagined a billowing cape of skin, the black-and-white diamonds of tissue in stark contrast with each other, the garment thrown around ORLAN's shoulders and worn with defiant provocation. Deflatingly, Oron admits that the work ended up being somewhat symbolic and largely abstract. From the biopsy that was taken of her skin and the tissue obtained from the ATCC, ORLAN, with the help of SymbioticA, was able to grow a very thin, almost-transparent layer incorporating both sets of skin cells in a special environment designed for the project.

It sounds a shadow of ORLAN's original vision of a black-and-white patterned membrane, a hybrid, 'a symbol of cultural crossbreeding'; but Oron defends the integrity of the work. While the realisation of ORLAN's original project may be largely symbolic, the cells of two different individuals were there, in the Petri dish.

'We know they are there,' he says, 'because we've done it, and that's very important for us ...' But for the audience, he concedes, 'it's really hard to figure out if what you see is the real thing. That's a constant issue, not just with ORLAN's work [but] with quite a lot of works that engage with this field. There are quite a few artists who are really good hoax artists. It is a very valid form of artistic strategy. It's really interesting when our work is being shown alongside some of those hoax artists'—they're so nice, they're so big, they're so amazing, so impressive, and ours are so ...' he pauses for a moment to find the right word, 'disappointing.'

CONTEMPLATING these artists' engagement with skin makes me want to confide my own squeamishness and unease to someone or something. Perhaps I could beg a small, semi-living worry doll, all skin-pink and mutely confessional, from Oron and Ionat. Whispering in its ear, though, my breath—pungent and ripe with bacteria and carrying any misgivings of skin, flesh, and the life within manipulated for art—would seed its death.

Besides, any qualms that I might have about these artworks are offset by my imaginings of delicate lace and hybrid harlequin coats spun from gossamer threads of skin.

The Skin We're In

> You never really understand a person until you consider things from his point of view … until you climb into his skin and walk around in it.
>
> — *To Kill a Mockingbird*, Harper Lee

DESPITE THE NATURAL ANTIPATHY between water and paper, I'm back in the Japanese bathhouse in the company of the women from my writing group for our regular get-together. Here we recline, another set of empresses in their new clothes. The popular image of authors doesn't include nudity so much; it's more one of shabbily dressed individuals hunching over their computers. Still, here we are, a bunch of writers in our birthday suits, sitting in the sunken tub, or perching on the surrounding bench when the deep penetrating heat of the water gets too much.

In the warmth and steam I have a sense of flesh, of softness, and of our bodies' wholeness that contrasts with the hard, cool tiles and the metallic shine of stainless-steel taps. Revealed is the uninterrupted fall of a smooth back, from rounded shoulder to the gentle swell of hips, the usually hidden tattoo, the small imperfections and blemishes wrought by sun, ageing and childbirth, all the tucks, dimples, and folds of our pliable hides.

Our skin is not on display, as such, but is simply the exterior of ourselves.

Here in this humid room, we see each other without judgement, without shame, comfortable in our own skins. Our collective nakedness makes me feel kinder and more tender towards us all.

On this night, there is a young Amazon in the bath with us, with her pubic hair carefully trimmed and shaped into a small, slim rectangle. I feel benevolently warm even towards her and, somehow, with all of us bare and exposed, don't even feel a twinge of envy for her smooth, young skin.

In the shadowiness of the bathhouse, our talk ebbs and flows in a gentle back and forth of conversation. We talk of writing, of friends who are ill, of travel, of theatre shows and movies that we have seen.

With each of the subjects that we discuss and topics that we touch on, our borders shift, our emotions and senses expand to absorb the experience of another. We briefly try on the skins of those who figure in the anecdotes and stories offered up, and of the speakers who frame the tales. It's a burden, but also a liberation; to eschew the position of objectivity and to slip into the skin of another—to attempt to feel their experiences as keenly as we feel our own, with all the intensity and clarity that they do. We live a little of the fear and anger of the woman who is ill, her sense of loss and outrage at the prospect of her life cut short, and we test the weight of the burden borne by the friend who cares for her and will grieve when she is gone. We reflect on the courage it takes to live in the face of such an illness and the valour to stand beside those who face their death.

It's not all grave discussions of mortality, however. We're able to temporarily colonise the usually decorous space of the bath with the weight of our numbers and, while there's no flicking each other with towels or boisterous splashing, there is gossip and laughter. Two of us, we discover, have separately been to see a recent performance of a play featuring Alison Whyte, she who covets a pair of boots fashioned from the skin of her fellow thespian.

As coincidence would have it, Marcus, he of the 'beautiful, dark silky skin' in which Alison quipped she'd like to see her own calves, feet, and ankles sheathed, was in the play, too. Alison was not naked as she strutted on the stage, but was not far from it, dressed only in a white, alarmingly cut-away swimsuit that served as a costume. We exclaim how buff she looked—'Her, a mother several times over and a woman in her forties!'

Marcus was equally revealed, and just as cut, in brief white bathers, his rippled abdomen and sculpted legs giving no indication that he was anywhere near ready for the knackers' yard. Alison might have a long wait before her well-toned pins can be wrapped in leather made from his silky pelt.

Pelt. It was a word I considered using as a concise and pithy title for this exploration of skin. It baldly declares the notion of the skin as separate or removed from the body; a thing in and of itself. Such a view imposes an emotional distance from which to view our remarkable rind, and is, in some respects, perverse. Thinking of the skin as a pelt not only grants permission to stroke and touch it, as if it were an object divorced from the person to whom it gives shape, but also to view it as a resource that can be harvested and put to use.

I think again of *The Poetical Works of Rogers, Campbell, J. Montgomery, Lamb, and Kirke White*, and ponder once more the provenance of that volume's binding. Did the sugary verse warrant the opportunistic utilisation of that sombre and grisly jacket?

The removed hide is also a reminder of what is left behind: the twitching, bleeding corpse; the muscle, bone, and sinew exposed. Something in pieces, no longer whole.

My visit to the knackers' yard provided me with a brutal demonstration of how an animal—and this includes the human animal—can be reduced to the sum of its parts; all it takes is the crack of a rifle shot and the skilful wielding of an icily keen blade to transform a creature into mere blood, flesh, bone, and skin.

It is something I have seen before; I grew up on a farm and I have witnessed the death of animals for profit and convenience, sometimes for sport and, occasionally, out of compassion. With these experiences under my belt, I was able to watch the casually dealt death at the knackers' yard not indifferently but dispassionately. The horses were despatched with a matter-of-fact bullet to the head. There may not have been much kindness in the act, but neither was there malice or deliberate cruelty. It was just the end of the line for animals that had been judged past their use-by date. I did not try to climb into the sad, pillaged skins lying on the wet concrete floor, still moist from the flesh of the animals that they had held together—although I did reflect on the backbreaking labour and small financial reward that was the lot of the hard and wiry men who did the work.

That fish-silver dawn spent at the knackers' yard brought home to me in a visceral and chilling way this notion of the skin as a pelt; something that could exist separately from body, but only with the death of the animal that it encased. Yet, at the end of this process of writing, this dissection, if you like, of the skin and of its meanings and its functions, it is not the separateness of the skin that has been reiterated for me again and again, but its interconnectedness with the body and the mind: the various ways that it reflects our inner emotional state, mediates our relationship with others, or broadcasts the state of our health. The more that I tried to detach the skin from its moorings, the tighter it cleaved.

Skin does so much more than simply define the point at which we begin; for while it is in essence a border, it is one that must allow traffic. It admits and repels, translates and transmits, shrinks and expands, receives and recoils. It is a flimsy boundary this skin of ours: thin and easily torn. It has to be to serve its purpose. If you clasp another to your breast you will feel the thud of their heart against your chest as if it were your own.

Skin is something that we all share and, as such, it reminds me of the things we have in common, and creates a space for empathy. For all that skin is as a site of difference—of race, of age, tribal allegiance—it is, more crucially, a site of connection. We all blush, sweat, feel, and bleed. We are revealed through our skin—the sweat streaks of sex, of labour, of endurance; the flush of shame, anger, arousal; the shiver of delight, of cold, of fear; the tear stains of frustration and despair; the creases of laughter, grief, and sadness; the furrows and sun-corrosion that betray our age.

The skin continually brings us back to the corporeality of ourselves and our physical places in the world. It is just a whisper-thin barrier between us and the cosmos, the stardust from which we are all made. Perhaps through the contemplation of our skin and its fragility, we can be reminded to handle each other with more care. And, at the same time, to see beyond the skin and the stereotypes that we may associate with tattoos, skin colour, wrinkles, and even youth and beauty.

Contemplating and scrutinising the skin has provided an opportunity to meditate on the life that we share, and a reminder of how thin and brittle is our grasp on life. It doesn't take much to tear the skin, or to wrench us away from the existence that we cling to so tenuously. My journey through skin has left me more accepting of the flaws of my own epidermis, more mindful of its vulnerability. I still look with envy on the unlined faces of the young and the miraculously preserved, but my own skin is a wonder in itself. I look at my own skin more keenly then ever, as it degenerates before me in the bathroom mirror—some days with equanimity, others with resigned regret.

As I've spent time considering my skin and its functions, and attempting to expand my awareness to take in every nuanced signal that it transmits, I've come to realise that one can't live forever in a constant state of hypersensitivity, at least not constructively or happily—noticing each trifle, each minor prick, each rise and drop of temperature by a fraction of a degree, each of the changing sensations induced by the sweep of wool or hard unyielding surfaces. It gets exhausting.

In some respects, then, it will be a relief to let go of this intense scrutiny of my hide, to simply allow it to fulfil its myriad functions as it has always done. But, hopefully, that sense of connection, not just with those whom I love but with all of those whom I inadvertently brush up against or manoeuvre carefully around, will remain.

BACK IN THE BATHHOUSE, flushed and rosy after our immersion, we sit in the lounge, my companions and I, clad in the loose cotton pyjamas that are provided by the establishment. We drink Japanese beer and snack on the house mix of wasabi peas, dried squid, and nuts. Several of us invest two dollars in the massage chair that further soothes our already-pampered bodies.

Despite the soft brush of the cloth now swaddling us, still radiating heat from the steam and the feverish water, we are all intensely aware of our skin.

Sources

Skin: the body's envelope

Information about the physiological aspects of skin in this chapter are drawn from *The Oxford Companion to the Body*, C. Blakemore and S. Jennett, (eds), Oxford University Press, Oxford, 2001; 'Skin', C. R. Robbins, in *Encyclopedia of Human Biology* (2nd edition), R. Dulbecco, (editor-in-chief), Academic Press, La Jolla, 1997; and *Anatomica: the complete reference to the human body and how it works*, D. Tracey, Random House, Sydney, 2000

My own observations and knowledge of the cultural and folkloric meanings and associations of skin have been augmented and expanded by *Skin: on the cultural border between self and the world*, C. Benthien, Thomas Dunlop (trans.), Columbia University Press, New York, 2002; *The Body's Edge: our cultural obsession with skin,* M. Lappé, H. Holt, New York, 1996; *Are You Superstitious?* L. Cowan, Leslie Frewin, London, 1968; and *Myth: its meaning and functions in ancient and other cultures*, G. S. Kirk, Cambridge University Press, London, 1970

p. 4: 'You gotta have skin …', from 'Skin' *Allan in Wonderland*, A. Sherman, Warner Bros, recorded 1964

p. 13: For detrimental effects on children and babies denied touch, see 'Touch Hunger', T. Field, *Touch*, MIT Press, Cambridge Massachusetts, 2001

pp. 13–14: For advantages of kangaroo care for preterm babies, see 'Kangaroo Mother Care', L. Albright, *Leaven*, Oct/Nov 2001, Vol. 37, 5,

p. 106; 'Touch and Massage' and 'Kangaroo Care', Parent–Infant Research Institute pamphlets, c2009; and 'Kangaroo Cuddles Help Premature Babies', S. Cauchi, *The Age*, 2 July 2002

p. 18: 'brought into relation with external objects …', see *Gray's Anatomy (Anatomy: Descriptive and Surgical)*, H. Gray, The Promotional Reprint Company Limited, Leicester 1991, [1858] p. 542

Touch Me

For details on touch and the nervous system, I have drawn from *The Oxford Companion to the Body*, Blakemore and Jennett, 2001; 'The "Sense of All Senses"' E. D. Harvey, in *Sensible Flesh: on touch in early modern culture*, University of Pennsylvania Press, Philadelphia, 2003; and *Touch*, T. Field, MIT Press, Cambridge Massachusetts, 2001

p. 20: 'She reminded him of the pleasure of being scratched …', from *The English Patient*, M. Ondaatje, Picador, London, 1993, p. 225

p. 31: For results of research into the advantages of massage for children and adolescents, see 'Massage Therapy Improves Mood and Behaviors of Students with Attention-Deficit/Hyperactivity Disorder', S. Khilnani, T. Field, M. Hernandez-Reif, and S. Schanberg in *Adolescence*, Winter, Vol. 38, 152, 2003, pp. 623–28

p. 31: 'Teenage sexual promiscuity and pregnancy are on the increase …', see *Touch*, T. Field, 2001, p. 5

p. 32: Conditions in Romanian orphanages following fall of Ceausescu regime described in 'My Experience in a Romanian Orphanage' G. Settle, *Massage Therapy Journal*, Fall, 1991, pp. 64–72

p. 37: For more information on the Royal Touch and *mal le roi*, see *The Royal Touch: sacred monarchy and scrofula in England and France*, M. Bloch (trans.

J. E. Anderson), Routledge and Kegan Paul and McGill, London, 1973

Melting Pot: the colour of skin

p. 43: 'I have a dream . . .', Martin Luther King, spoken in 1963, transcribed at <www.usconstitution.net/dream.html>

p. 46: *The Book of Enoch*, R. H. Charles, (ed.), Oxford, Clarendon, 1893

p. 47: *The Da Vinci Code*, D. Brown, New York, Doubleday, 2003

p. 47: Information on albinism drawn from 'Albinism—Awareness and Understanding: a handbook for anyone interested in the welfare of people with albinism', J. Sullivan, Albinism Fellowship and Support Group Inc., Adelaide, 1998; and the National Organization for Albinism and Hypopigmentation <www.albinism.org>

pp. 50–51: 'If you are brown and decide to date a British man . . .', from 'You are in Paradise', Z. Smith, the *New Yorker,* Vol. 80, 16, 14 June 2004, p. 150

p. 51: Article published in the *Scotsman* about national skin-colour survey, see 'Birthplace more than skin colour makes a "real" Scot', W. Lyons, the *Scotsman*, 6 February 2004

p. 52: 'His view was that it was the superiority . . .' *The Characters of the Human Skin in Their Relations to Questions of Race and Health*, H. J. Fleure, Oxford University Press, London, 1927, p. 28

pp. 52–53: 'He suggested the compromise of a "sheltered life" . . . ibid, p. 30

pp. 53–54: Role of melanin and other factors in determining skin colour drawn from *The Oxford Companion to the Body*, Blakemore and Jennett, 2001; and *Our Kind: who we are, where we came from, where we are going*, M. Harris, Harper & Row, New York, 1989 pp. 55–58: Research into relationship between UV, vitamin D, folate, and skin colour published in 'Skin Deep', Nina G. Jablonksi, and G. Chaplin, in *Scientific American*, Vol. 13, 2, July 2003, pp. 72–80; 'Why Skin Comes in Colours', B. Edgar, *California Wild*, Vol. 53, 1, Winter, 2000, pp. 6–7; and 'Black and White', *Discover* Vol. 22, 2, February 2001

p. 55: 'The *Medical Journal of Australia* identified a new high-risk …', see 'Vitamin D deficiency and multicultural Australia', R. S. Mason and T. H. Diamond, 2001, *Medical Journal of Australia*, No. 175, pp. 236–37

p. 58: James Mackintosh's research, see 'Perfect skin: In the tropics, darker skin could well be the best health insurance', *New Scientist*, N. Jones, Vol. 170, 2288, 28 April 2001; 'white-skinned soldiers were three times more likely …', see 'New theory on why dark skin protects in the tropics', Australian Broadcasting Commission,<www.abc.net.au/science/news/stories/s282901.html>

p. 59: Predictions of Dr Oliver Curry, see 'Welcome to the year 3006 and—if we're still alive—a brave new world where man is tall, dark and incredibly handsome', N. Fleming, *The Age*, 18 October, 2006, p. 3

O, No! It is an Ever-fix'd Mark: scars, moles, and other blemishes

p. 60: 'And the Lord said unto him …', from Genesis 4:14–16

p. 62: *A Wizard of Earthsea*, U. Le Guin, USA, Parnassus Press, 1968

pp. 64–65: For descriptions and photos of Nuba scarring practices see *Nuba Personal Art*, J. C. Faris, University of Toronto Press, Toronto, 1972

p. 64: For information on the process of scarring, I referred to *The Oxford Companion to the Body*, Blakemore and Jennett, 2001 pp. 67–68: Didier Anzieu on mutilations of the skin, see *The Skin Ego: a psychoanalytical approach to the self*, D. Anzieu, (trans. C. Turner), Yale University Press, New Haven and London, 1989, p. 20

p. 70: Newspaper report of 2002 Bali bombings, see 'Tracing a killer smile', D. Goodsir, *A3, The Age*, 6 August 2003, p. 7

p. 73: Biblical account of the mark of Cain found in Genesis 4:14–16

p. 73–74: For more information on Cain and his descendents, see *Who's Who in the Old Testament Together with the Apocrypha*, J. Comay, Werdenfeld, and Nicholson, London, 1971, and *Companion to Literary Myths, Heroes and Archetypes*, P. Brunel, (ed.), (trans. W. Allatson, J. Hayward, T. Selous), Routledge, London, New York, 1992

p. 76: For more on superstition and folklore relating to marks and scars, see *Dictionary of Mythology, Folklore, and Symbols*, G. Jobes, Scarecrow Press, New York, 1961–62; *Are you superstitious?*, L. Cowan, Leslie Frewin, London, 1968.

p. 76: 'Distrust those marked by the Creator …', from *Lady's Pictorial*, 21 November 1914, p. 713

p. 77: Cases of children with scars that reportedly correspond with injuries suffered in past lives found in *Twenty Cases Suggestive of Reincarnation*, I. Stevenson, American Society for Psychical Research, New York, 1966

pp. 78–79: For more information on the Devil's mark and witch-prickers, see *Satanism Today: an encyclopedia of religion, folklore, and popular culture*, J. R. Lewis, ABC-Clio, Santa Barbara, California, 2001

pp. 79–80: Popular perceptions of the 'beast', see *The Oxford Companion to the Bible*, D. H. van Daalen, Oxford University Press, B. M. Metzger and M. D. Coogan, (eds.), Oxford, 1993, p. 700

pp. 80–81: Information on the history of makeup and the decorated body comes from *The Artificial Face*, D. and C. Fenja Gunn, Newton Abbot, London, 1973; *A history of make-up* M. Angelou, Studio Vista, London, 1970; *The Decorated Body*, R. Brain, Hutchinson & Co., London, 1979; *The Body Decorated*, V. Ebin, Thames and Hudson, London, 1979; and *The Gilded Lily*, T. McLaughlin, Cassell & Company, London, 1972

p. 81: Article about the lady at liberty to wear a patch on the side she pleases, see *The Spectator*, 2 June 1711

Peculiar to Humanity: blushing and tickling

For detail on the physical process of blushing, I drew from *The Oxford Companion to the Body*, Blakemore and Jennett, 2001

p. 84: '[blushing] is the color of virtue.', from *Diogenes of Sinope: the man in the tub*, Luis E. Nava, Greenwood Press, Westport Connecticut, 1998, p. 162

p. 86 'As beautiful as a woman's blush …' From 'Apple Blossoms', L. E. Landon, *Hoyts New Cyclopedia of Practical Quotations*, J. K. Hoyt, Funk & Wagnalls Company, New York, London, 1922

p. 86: 'Blushing is the most peculiar and the most human of all expressions,' from *Expressions of the Emotions in Man and Animals*, C. Darwin, T. Murray, London, The University of Chicago Press, Chicago & London, 1965, [1872], p. 309

p. 86: 'Man is the only animal that blushes. Or needs to', from *Following the Equator: a journey around the world vol. i*, Mark Twain, The Ecco Press, New Jersey, 1992, [1897], p. 216

p. 86: Darwin's theory of emotions and blushing, see *Expressions of the Emotions in Man and Animals*, C. Darwin, 1965, pp. 309–10

p. 89: 'It is only in white men ...'. see A. du Humboldt and A. Bonpland, *Personal narrative of travels to the equinoctial regions of the New continent during the years 1799–1804'*, translated into English by H. M. Williams, Longman, Hurst, Rees, Orme, and Brown, London, 1814, p. 229

p. 93: 'In order that the soul might have sovereign ...', from *Expressions of the Emotions in Man and Animals*, C, Darwin, 1965, p. 336

pp. 95–100: Information about different types of tickling responses comes from 'The Mystery of Ticklish Laughter', C. Harris, *American Scientist*, Vol. 87, 1999, and from 'Tickling', S. Blakemore in *The Oxford Companion to the Body*, Blakemore and Jennett, 2001

p. 95: 'If you tickle us, do we not laugh?' W. Shakespeare, *The Merchant of Venice*, Act III, Scene I

Unclean, Unclean: skin and disease

Information about skin disease in this chapter, including leprosy, has been drawn from *The Encyclopaedia of Skin and Skin Disorders*, (3rd ed.), C. Turkington and J. S. Dover, Facts on File, New York, 2002; Lepra <www.lepra.org.uk/home.asp>; and DermNet <www.dermnetnz.org>

Information on psychocutaneous disorders drawn from 'Recognising psychocutaneous disorders in general practice', L. Spelman and M. Spelman, *Modern Medicine*, Vol. 40, 5, May, 1997; 'Psychodermatology: the mind and skin connection', J. Koo and A. Lebwhol, in *American Family Physician*, Vol. 64, 11, 1 December 2001; and 'Recognizing the Mind-Skin Connection', *Harvard Women's Health Watch*, November 2006

p. 101: 'The skin, after all is extremely *personal*, is it not?', from *The Singing Detective*, Dennis Potter, Faber and Faber, London, 1991, p. 56

p. 107: '... patients should no longer be referred to as "lepers".' from *The*

Encyclopaedia of Skin and Skin Disorders, C. Turkington, and J. S Dover, 2002, p. 211

p. 108: Dennis Potter speaks about his skin disease in *Potter on Potter*, Graham Fuller, (ed.) Faber, London, 1993

p. 110: Andrew Strathern on skin condition and semen, see *Body Thoughts*, A. J. Strathern, University of Michigan Press, Ann Arbour, 1996, p. 85

She's Got Perfect Skin

Information on skin, ageing, and efficacy of 'cosmeceuticals' drawn from DermNet NZ <www.dermnetnz.org>; *The Oxford Companion to the Body*, Blakemore and Jennett, 2001; and Turkington and Dover, *Skin Deep: an A–Z of skin disorders, treatments, and health*, Facts on File, New York, 1996

Historical information on beauty treatments found in *Fashion in Makeup from Ancient to Modern Times*, R. Corson, Universe Books, New York, 1972

p. 117: 'Louise is the girl with the perfect skin …', from: 'Perfect Skin', L. Cole, *Rattlesnakes*, Geffen Records, released 1984

p. 120: *Guardian* article on the harvesting of skin by a Chinese pharmaceutical company, see 'The beauty products from the skin of executed Chinese prisoners', I. Cobain and A. Luck, *Guardian* www.guardian.co.uk/print/0,3858,5284064-1088142,00.html, 13 September 2005

The Lure of the Tattoo

Images of tattoos referred to, and information on tattoos, drawn from: 'Tattooing Among the Arabs in Iraq,' W. Smeaton, *American Anthropologist*, Issue 39, 1, 1937; *Decorated Skin: a world survey of body art*, K. Gröning, Thames and Hudson Ltd, London, 1997; *Tattoo History, a source book: an anthology of historical records of tattooing throughout the world*, S. Gilbert, (ed.),

Juno Books, New York, 2000; and *Bodies of Subversion: a secret history of women and tattoo*, M. Miffin, Juno Books, New York, 1997

p. 132: 'I always look for a woman who has a tattoo …' R. Jeni, quoted by A. Blimes in 'What men really think about women and tattoos', *MailOnline*, 28 January 2008 <*www.dailymail.co.uk/femail/article-510827/What-men-REALLY-think-women-tattoos.html*>

Alison Whyte's Boots: skinning and the human pelt

p. 150: 'Tan me hide …' R. Harris, 'Tie Me Kangaroo Down, Sport,' *Tie Me Kangaroo Down, Sport* EMI, recorded 1963

p. 151: 'beautiful, dark, silky skin …', from 'Alison in Wonderland', B. Hooks, in *Sunday*, 3 August 1997

p. 152: The satyr in *Metamorphoses*, see *Metamorphoses*, Ovid, trans. A. D. Melville, introduction and notes by E. J. Kenney, Oxford University Press, Oxford, 1986, p. 133

p. 153: The story of Sisamnes found in *The History of Herodotus, Book V*, Herotodus, (trans. by G. Rawlinson), J. M. Dent & Sons, London, 1992

p. 155: *The Diary of Samuel Pepys Vol II, 1661*, S. Pepys, R. Latham, and W. Matthews (eds), G. Bell and Sons Ltd, London, 1971, p. 70

pp. 155–56: Tyack's essay, see 'Human Skin on Church Doors', Reverend G. S. Tyack, *The Church Treasury of History, Custom, Folk-lore, Etc*, William Andrews & Co, London, 1898, p. 159

p. 156: Information on identifying origin of the 'skins of the Danes' drawn from the BBC history website 'Viking Dig Reports 9. Hadstock Church', <www.bbc.co.uk/history/ancient/vikings/dig_reports_09.shtml>

p. 157: Information on the treatment of the corpses of executed criminals in 18th-century England was drawn from *Death, Dissection, and the Destitute*, R. Richardson, Routledge & Kegan Paul, London; New York, 1987

pp. 157–158: Reports of efforts to stamp out human skinning in Tanzania found on BBC News website 'Tanzania fights human skinning', <http://news.bbc.co.uk/go/pr/fr/-2/hi/Africa/3045738.stm> 4 July 2003; and 'Tanzania arrests 'witch killers', <http://news.bbc.co.uk/go/pr/fr/-2/hi/Africa/3209047.stm> 23 October 2003

p. 158: 'Breaking the Mirror: from the Aztec feast of the flaying of men to organ transplantation,' I. Clendinnen, 2002, *The Best Australian Essays 2003*, P. Craven (ed.) Black Inc., Melbourne 2003, p. 252

p. 158: *Aztecs: an interpretation*, I. Clendinnen, Cambridge University Press, Cambridge; New York, 1991

p. 159: Information on Eddie Gein drawn from 'Buffalo Bill & Psycho', R. Bell and M. Bardsley, at Crime Library website <www.crimelibrary.com/gein.geinmain.htm>

p. 160: '… him I like especially well …', from *The Shrine of Jeffrey Dahmer*, B. Masters, Hodder & Stoughton, London, 1993, p. 107

p. 160: '"pleasureable" sexual intercourse', see 'Life for "evil" killer', G. Wendt, in *Newcastle Herald*, 9 November 2001

pp. 160–61: Information on the murder of John Price by Katherine Knight also drawn from 'Gruesome taste in the video rack', G. Wendt, *Newcastle Herald*, 24 October 2001; and 'Thrill killer would kill again: court', G. Wendt, *Newcastle Herald*, 27 October 2001

p. 161: Interviews with prisoners of Buchenwald, see *The Buchenwald Report*, D. A. Hacket, (trans. & ed.), Westview Press, Boulder, 1995, p. 338

p. 162: Carl Whittaker, see CNN.com, <http//:cnn.worldnews…/

cpt?action=cpt&expire=-1&urlID=5521476&fb=Y&partnered=200>

pp. 162–63: News reports of the sale of Wim Delvoye's living tattoo were published on the Artinfo website: 'Artist Sells "Living Tattoo" on Another Man's Back' www.artinfo.com/news/story/28494/artist-sells-living-tattoo-on-another-mans-back, and in the *Courier Mail*, 'Man sells tattoo on his back to German collector for $257,880,' 2 September 2009

p. 163: Ingrid Newkirk's instructions regarding her will, see *Courier Mail*, 'Activist cooks up ultimate protest,' 16 April 2003

Bound in Human Skin: anthropodermic bibliopegy

p. 164: '... I find the application of this sort of leather ...', from *The Anatomy of Bibliomania*, H. Jackson, Faber and Faber, London, 1950, p. 404

p. 165: Volume on virginity, pregnancy, and childbirth bound in human skin, see *De integritatis & corruptionis virginum notis*, S. Pinaeus, L. Bonacciuoli, F. Platter, and M. Sebizii. Notes on the Wellcome Library's online catalogue entry for the book read: 'Human skin bound by Lortic ... Mentioned by Paul Combes in the journal "Intermédiaire des chercherus et curieux" in 1910, with a note that Bouland had acquired the piece of skin while a medical student, from the body of woman who had died in the hospital at Metz.'

p. 165: *The Dance of Death: from the original designs of Hans Holbein*, H. Holbein, printed for J. Coxhead, London, 1816

p. 165: *The Chronicles* is kept at the Newberry Library, Chicago. The inscription claims that it is 'a manuscript written in the year of the Hijrah 1266 (1848 AD)'.

p. 165: *Narrative of the life of James Allen: alias George Walton, alias Jonas Pierce, alias James H. York, alias Burley Grove the highwayman: being his death-*

bed confession, to the warden of the Massachusetts state prison, James Allen, Harrington & Co, 1837. Notes in the Boston Athenaeum's online catalogue entry for the book read: 'Bound by Peter Low in Allen's skin, treated to look like gray deer skin; bears the cover title "Hic liber Waltonis cute compactus est," stamped in gold upon a black leather rectangle.'

p. 166: Milton's works, see *The Poetical Works of John Milton*, J. Milton, William Tegg & Co, London, 1852. Notes in the Devon library's online catalogue entry for the book read: 'Bound in skin of George Cudmore, hanged for murder, 1830'.

p. 166: Report of skin-bound ledger found in Leeds published on the BBC News website: 'Police Plea on Macabre Book Find' <http://news.bbc.co.uk/2/hi/uk_news/england/west_yorkshire/4891100.stm>

p. 166: Jackson's Russian poet example, see *The Anatomy of Bibliomania*, H. Jackson, 1950, p. 405

p. 168: The account of Corder's trial bound in his own skin is held at the Moyse's Hall Museum. Further details of the crime and other exhibits relating to the Red Barn Muder can be found on the Moyse's Hall Museum website <www.stedmundsbury.gov.uk/sebc/visit/exhibits.cfm>

p. 168: *The Poetical Works of Rogers, Campbell, J. Montgomery, Lamb, and Kirke White*, S. Rogers, T. Campbell, J. Montgomery, C. Lamb, and H. K. White, A. and W. Galignani, Paris, 1829

p. 169: *Anatomia Completa del Hombre*, Martin Martinez, 1757

p. 171: Edwin Zaehnsdorf on turning human hide into leather, see *The Anatomy of Bibliomania*, H. Jackson, 1950, p. 403

pp. 172–73: For information on the dispute over Corder's remains, I drew from articles on the Culture 24 website, including: 'Suffolk Museum may return murderer's grisly remains to relative', G. Spicer

<www.culture24.org.uk/places+to+go/london/art44179>

p. 175: *My Life with Paper*, D. Hunter, Alfred Knopf, New York, 1958, pp. 41–2. For further information on anthropodermic bibliopegy, see *Bookmen's Bedlam: an olio of literary oddities*, W. H. Blumenthal, Rutger University Press, New Jersey; New Brunswick, 1955; and *Bibliologia Comica or Humorous Aspects of the Caparisoning and Conservation of Books*, L. S. Thompson, Acheron Books, 1968

You're Nicked: the romance of fingerprints

For the history and technical details of fingerprinting, I have drawn on: *Fingerprints: the origins of crime detection and the murder case that launched forensic science*, C. Beavan, Hyperion, New York, 2001; *Encyclopedia of Forensic Science*, S. Bell, Facts on File, New York, 2008; *The Forensic Casebook: the science of crime scene investigation*, N. E. Genge, Ebury Press, London, 2004; and *Encyclopedia of Crime Scene Investigation*, M. Newton, Facts on File, New York, 2008

p. 177: 'Every human being carries with him from his cradle to his grave …', from *Pudd'nhead Wilson: a tale*, Mark Twain, Chatto & Windus, London, 1894, p. 231

p. 177: 'Even the smallest of thumb screws was forbidden …' , from: the foreword to *Fingerprints: history, law, and romance*, G. W. Wilton, William Hodge and Company Limited, London, 1938, p. xi

p. 179: Somewhere about 1878 …', from *Fingerprints: history, law, and romance*, G. W. Wilton, 1938, p. 1

p. 179: 'When bloody finger–marks …', from 'The Skin Furrows of the Hand', H. Faulds, *Nature,* Vol. XXII, 28 October 1880, p. 605

p. 180: 'repudiating his signature…', from *The Origin of Finger-Printing*, W.

Herschel, Oxford University Press, London, 1916, p. 8

p. 180: Sir William Herschel's use of fingerprints, see W. Herschel, 'Skin Furrows of the Hand' in *Nature*, Vol. XXIII, 25 November 1880, p. 76

p. 181: Twain on fingerprints as an autograph, see 'These marks are his signature, his physiological autography…', from *Pudd'nhead Wilson: a tale*, M. Twain, 1894, p. 231

p. 181: 'The Adventure of the Norwood Builder', A. C. Doyle, *The Return of Sherlock Holmes*, found at Project Gutenberg, <www.gutenberg.org/files/108/108-h/108-h.htm#2H_4_0002, 2006>

pp. 187–88: Information on Shirley McKie's case drawn from news.scotsman.com: 'Shirley McKie fingerprint inquiry opens' <http://news.scotsman.com/shirleymckiefingerprintcase/Shirley-McKie-fingerprint-inquiry-opens.5324450.jp>; and 'Shirley Mckie fingerprint case' <http//:news.scotsman.com/newsfront.aspx?sectionid=8493&IsTopic=1>, as well as shirleymckie.com <www.shirleymckie.com/index.htm>

p. 188: *New York Times* report on flawed identification, see 'Report Faults F.B.I.'s Fingerprint Scrutiny in Arrest of Lawyer', D. Stout, *New York Times*, 17 November 2004

Peering into the Abyss: burns injuries

p. 192: 'Burns are also a kind of an abyss to peer into …', *Rising from the Flames: the experience of the severely burnt*, A. H. Carter III and J. A. Petro, University of Pennsylvania Press, Philadelphia, 1998, p. 29

p. 193: Story of Prometheus and Epimetheus, from *Bulfinch's Mythology*, T. Bulfinch, The Modern Library, New York, c1982

p. 195: Research on the immediate impact of the Back Saturday fires, from

'Black Saturday: the immediate impact of the February 2009 bushfires in Victoria, Australia', P. A. Cameron, B. Mitra, M. Fitzgerald, et al, *Medical Journal of Australia* Vol. 191, 1, 6 July 2009, pp. 11–16

p. 214: *Total Burn Care* (3rd edition), D. N. Herndon, Saunders Elsevier, 2007

Leaves on the Tree of Life: the Donor Tissue Bank of Victoria

p. 222: 'The only gift is a portion …', from 'Gifts', Ralph Waldo Emerson, in *The Complete Works of Ralph Waldo Emerson Vol. 1*, Bell and Daldy, London, 1866, p. 221

Second Skins: semi-living sculpture

p. 233: 'Garments are humans' fabrication …', from 'Victimless Leather—A Prototype of Stitch-less Jacket grown in a Technoscientific "Body"', Tissue Culture and Art Project, <www.tca.uwa.edu.au/vl/vl.html>

p. 241: 'Museum Kills Live Exhibit', J. Schwartz, *The New York Times*, 13 May 2008

p. 243: 'the gap created by the slow and non-purposive biological evolution …', from Tissue Culture and Art Project, <www.tca.uwa.edu.au/project/overview/over_home.html>

More information on Symbiotica, Tissue Culture and Art Project, and ORLAN can be found on their respective websites: <www.symbiotica.uwa.edu.au>; <www.tca.uwa.edu.au>; <www.orlan.net>

The Skin We're In

p. 249: 'You never really understand a person …', from *To Kill a Mockingbird*, H. Lee, Pan Books, London 1976, p. 35

Acknowledgements

I have taxed the patience and goodwill of many in researching this book, and I am deeply indebted to many for sharing their experience, expertise, and opinions with me.

My thanks and gratitude goes to:

Wendy Meddings, Philippa Frances, Carol Newnham, Shelley Robertson, Duncan Macgregor, Janet Davey, Pam Craig, Anne Davies, Jenny Webb, Josie Yeatman, Kenny McPharlane, Marisa Herson, Kellie Hamilton, and Steven Dunn for their personal and professional insights.

Elizabeth Caplice, for alerting me to the existence of a skin-bound book at the NLA, and to Pam Pryde, for drawing my attention to *Anatomia Completa del Hombre.*

Stephen Taberner, Maryjeanne Watt, and Jo Clifford for stories about their skin.

Grazi Lisciotto and Karen Moran for their generosity and openness in speaking about their experiences.

Bev Nicholas and Tim Dywelska for letting me sit and watch while they went about their work, and for answering my questions while they did so.

Peter Benda, Emily (the 'young lady' with the flamingo tattoo), my sisters Christine and Madonna, and Toni Moynihan (whose turtle may not have brought her longevity, but whose goal to have a fertile and creative life was certainly achieved), for showing me their tattoos.

Graham Kent and Deb Withers, for pointing me in the right direction, and for smoothing the way.

Fiona Wood and Yvonne Singer, for sharing their passion, time, and knowledge.

Oron Catts and Ionat Zurr, for a fascinating few hours discussing art and the semi-living.

Nicola Redhouse, my editor, for her patience and thoroughness.

And, finally, to the stalwart women of my writing group, Sam, Michelle, Myf, Wendy, Jane, and Spiri: your encouragement, support, criticism (always constructive), and friendship have been, and continue to be, invaluable to me.